The Logic of the Third

A Paradigm Shift to a Shared Future for Humanity

World Scientific Series in Information Studies
(ISSN: 1793-7876)

Published:

Vol. 14 *The Logic of the Third: A Paradigm Shift to a Shared Future for Humanity*
by Wolfgang Hofkirchner

Vol. 13 *Ontological Information: Information in the Physical World*
by Roman Krzanowski

Vol. 12 *Trilogy of Numbers and Arithmetic*
Book 1: History of Numbers and Arithmetic: An Information Perspective
by Mark Burgin

Vol. 11 *Theoretical Information Studies: Information in the World*
edited by Mark Burgin & Gordana Dodig-Crnkovic

Vol. 10 *Philosophy and Methodology of Information: The Study of Information in the Transdisciplinary Perspective*
edited by Gordana Dodig-Crnkovic & Mark Burgin

Vol. 9 *Information Studies and the Quest for Transdisciplinarity Unity through Diversity*
edited by Mark Burgin & Wolfgang Hofkirchner

Vol. 8 *The Future Information Society: Social and Technological Problems*
edited by Wolfgang Hofkirchner & Mark Burgin

Vol. 7 *Information Theory Models of Instabilities in Critical Systems*
by Rodrick Wallace

More information on this series can also be found at https://www.worldscientific.com/series/wssis

World Scientific Series in Information Studies — **Vol. 14**

THE LOGIC OF THE THIRD

A Paradigm Shift to a Shared Future for Humanity

Wolfgang Hofkirchner
The Institute for a Global Sustainable Information Society, Austria

NEW JERSEY • LONDON • SINGAPORE • BEIJING • SHANGHAI • HONG KONG • TAIPEI • CHENNAI • TOKYO

Published by

World Scientific Publishing Co. Pte. Ltd.
5 Toh Tuck Link, Singapore 596224
USA office: 27 Warren Street, Suite 401-402, Hackensack, NJ 07601
UK office: 57 Shelton Street, Covent Garden, London WC2H 9HE

Library of Congress Control Number: 2022047095

British Library Cataloguing-in-Publication Data
A catalogue record for this book is available from the British Library.

World Scientific Series in Information Studies — Vol. 14
THE LOGIC OF THE THIRD
A Paradigm Shift to a Shared Future for Humanity

ISBN 978-981-126-101-5 (hardcover)
ISBN 978-981-126-102-2 (ebook for institutions)
ISBN 978-981-126-103-9 (ebook for individuals)

For any available supplementary material, please visit
https://www.worldscientific.com/worldscibooks/10.1142/12985#t=suppl

Typeset by Stallion Press
Email: enquiries@stallionpress.com

Printed in Singapore

for Gerti

Preface

This book is a sequel to my book "Emergent Information". Both contributions seek to highlight a paradigm shift as profound as any in the history of humankind. This shift is in the making and would boost scientific and everyday thinking onto a new level if fully realised: it requires nothing short of a radical revolution to master the global challenges that are existential threats to the thriving and survival of the human race. The key to participating in a trajectory that promises to prolong the social evolution of humanity is to acknowledge that systems long for what I call the Logic of the Third. This refers to two leaps. The first is a real leap onto the level of co-operation within one common overarching system, which is to be established as soon as the complexity of challenges exceeds the complexity of singular non-co-operative systems. The second is an ideational leap onto a meta-level of knowledge, which is to be newly-created by human informational agents in anticipation and orientation for the first, real leap.

The first book discussed three fundamental building blocks of the new paradigm, that is, Praxio-Onto-Epistemology (POE) as a philosophy, an Evolutionary Systems Theory (EST) based on POE, and a Unified Theory of Information (UTI) based on EST. The second book brings these blocks to fruition in the context of the current "social", social information and social information technology and elaborates on a framework of three more concrete building blocks: a Critical Social Systems Theory (CSST) based on EST, a Critical Information Society Theory (CIST) based upon CSST and UTI, and a Critical Techno-social systems Design Theory (CTDT) based upon CIST.

This approach contemplates a Science of Transformation that is necessary to scientifically base future political decisions designed to implement a Great Transformation – the techno-eco-social transformation into a Global Sustainable Information Society. That society takes on the role of the real and concrete utopia today. Departing from a mathematical term, transformation has become a term whose meaning has been eclipsing and contextualising any other problem worth solving. Hence transformative imperatives for a shared humanity.

As with the first book, I focussed on consistency of an integrated approach, not on lengthy matters of contention. My goal was to keep the text concise while adding tables to explicate the red thread and figures to illustrate the meaning of theoretical thoughts. I formulate methodological principles that should be helpful for further research in the wake of the new paradigm.

Going beyond the first book was an adventure equally joyful and fascinating as writing that book. I owe thanks to many new as well as old peers and friends for eye-opening discussions. I learned much from scholars in the field of social and human sciences whom I had the chance to get to know. Let me mention my several-years-long collaboration with Margaret S. Archer, Pierpaolo Donati, Douglas V. Porpora, Colin Wight, Philip S. Gorski, Tony Lawson, Emmanuel Lazega and Jamie Morgan. I connected with colleagues elaborating convivialist ideas, in particular Edgar Morin, Wilfried Graf, Frank Adloff and Werner Wintersteiner, and I had an intensive discussion with Chantal Mouffe. Though I have known him from literature since the days of my studies, I only met philosopher of science Mario Bunge personally in the last decade and I am grateful to him for his straightforward support of my research endeavours. He passed away at the age of 100 in 2020. I met Hans-Jörg Kreowski, with whom I enjoy a productive collaboration in critical perspectives on computer science. I am thankful to Peter Crowley for his expertise relating to the UN family when dealing with ICTs for global citizens, global dialogue and global governance. I was able to share ideas with Sarah Spiekermann(-Hoff) with regard to the ethical alignment of ICTs. With regard to the study of information, I am thankful for stable collegial relationships with Kun Wu, Yixin Zhong, Mark Burgin, Gordana Dodig-Crnkovic, Joseph Brenner and Kang Ouyang just to name a few.

Of course, I have remained in touch with colleagues from different fields whose acquaintance was already important for my first book, among them Rainer E. Zimmermann, José María Díaz Nafría, Klaus Kornwachs, Rafael Capurro, Robert K. Logan, Christian Stary, Pedro C. Marijuán, Yagmur Denizhan, Iryna Dobronravova, Yurii Mielkov and Dail Doucette. I regret very much the death of John Collier in 2018, with whom I had a years-long exchange and co-operation. Finally, I thank Michael Stachowitsch for tightening up the English text. This list is, of course, not exhaustive.

The message of the book is this: scientific rigour can manifest humanity's underlying disposition to usher in a new, third step in anthroposociogenesis and to correct the outdated logics that has hampered its unfolding to date. It is imperative to roll back imperial intentions, intransigent interests and idiotist identities. And it is entirely reasonable to assume that this lies within the human potential.

Wolfgang Hofkirchner, Vienna, March 2022

Contents

List of Tables

List of Figures

Part I

Towards a Science of Transformation

Chapter 1

Revisiting the World, Systems and Information

> Change is regular *and* irregular; consistent *and* inconsistent; continuous *and* discontinuous. Since the only logics available have been propositional bivalent logics, incapable of accepting the real contradictions present or implied in the description of real phenomena, they have been incapable of describing change.
>
> *– Joseph E. Brenner, Abir U. Igamberdiev: Philosophy in Reality, A New Book of Changes, 2021 –*

This book is a logical extension of the book *Emergent Information – A Unified Theory of Information Framework*. The scientific content of that book provides the building blocks of a paradigm shift in three domains, namely, firstly, in philosophy towards emergentism (Praxio-Onto-Epistemology – POE), secondly, in systems thinking towards emergentist systemism (Evolutionary Systems Theory – EST) and, thirdly, in the study of information towards emergentist-systemist informationism (Unified Theory of Information – UTI), whereby each step builds upon the previous one. The current book presents three further steps: from cornerstones of EST to cornerstones of a Critical Social Systems Theory; from cornerstones of the UTI and the Critical Social Systems Theory to those of a Critical Information Society Theory; and from the latter to cornerstones of a Critical Techno-Social Systems Design Theory. The term "critical" heralds the extension of the paradigm shift to social sciences, to science, technology and society studies and further to technology assessment, technology design and informatics, to name the main disciplines affected. Inspired, as it is, by ideas of the Critical Theory, criticism means putting one's own discipline in question, reflecting upon its practices, foundations and methods, and thus distancing the discipline from itself so that it reflects upon itself from a higher point of view. This is the scientific approach at its best and helps avoid falling prey to prejudices of social origin. The general principles for theorising the world as a world of unity-through-diversity, as a world of systems and as a world of systems that generate information are developed into more specific principles for theorising the social world, the informational social world and an informational social world that is technologically connected.

Together, those principles form a theoretical framework that goes beyond a mere mental exercise with no real-world implications. This is the proper basis for elaborating successful practical responses to the global challenges that pose an existential threat to humanity. From the perspective of the framework provided here, those challenges represent a crisis in the coming-of-age of humanity that can be overcome by taking the next step in social evolution. That framework characterises the present situation as a Great Bifurcation between a breakthrough to a new social systems trajectory and a breakdown of the old one with no substitute. A transformation of society, ecology and technology – a techno-eco-social transformation – is needed to master the leap to a Global Sustainable Information Society. Implementing this upgraded vision of the "good society" means ensuring the survival and thriving of all humanity by mindfully containing self-inflicted disruptions in the social organisation of the relations among humans, between humans and nature and between humans and technology. This book seeks to unfold an understanding of what the imperatives of our time are and how humans can comply with them. This effort will have a better social impact when on a sound foundation, which was provided by the first book – *Emergent Information – A Unified Theory of Information Framework.*

The paradigm shift to systems and informational systems in its new emergentist understanding is a shift away from the traditional edifice of sciences (Figure 1.1). In that overview, philosophy adopts the most general position. This is juxtaposed to the three categories of formal sciences (e.g. logics and mathematics), real-world sciences (divided into the natural sciences versus the social and human sciences) and applied sciences (e.g. engineering, management or arts) along with its subdivisions (e.g. physics, chemistry, biology, or sociology). All are siloed against each other by impermeable boundaries. A connection between those mono-disciplines can be attempted only by heaping some of them together in a multi-disciplinary approach, which is actually no connection at all. Another strategy would be to envision peripheral exchanges in an interdisciplinary approach. That, however, does not admit internal changes and fails to bridge the traditional alienation between the disciplines. Informatics, social informatics, informatics and society are, like many other sub-subdivisions, even more devoid of a clear positioning in that edifice and struggle to find a clear identity.

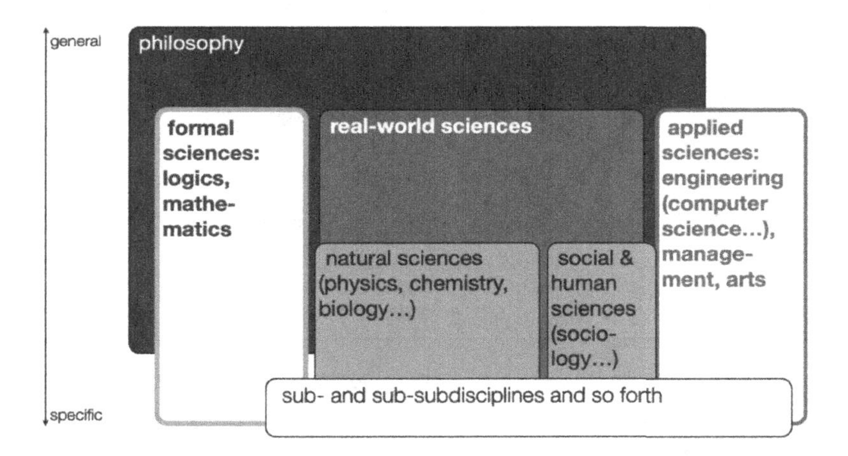

Figure 1.1. Edifice of sciences. Normal science.

Addressing issues of complexity, which is unavoidable, clearly manifests the deficits of this understanding of the science edifice. Taking the systems perspective helps alter our comprehension of the edifice (Figure 1.2):

- philosophy, which was deprived of fruitful relations to the disciplines of science in what had become "normal science" in Kuhn's diction [Kuhn 1962], now turns into systems philosophy;
- formal sciences turn into formal as well as non-formal systems methodology; real-world sciences turn into sciences of real-world systems, that is, material, living or social systems (tenets of social systems science are elaborated in Part II from a point of view that

is critical of society – a Critical Social Systems Theory – to distinguish it from social systems approaches, which do not pertain to the new paradigm);

- applied sciences turn into a science that creates artefacts by designing systems and, in doing so, integrates them with social systems (this also involves criticism).

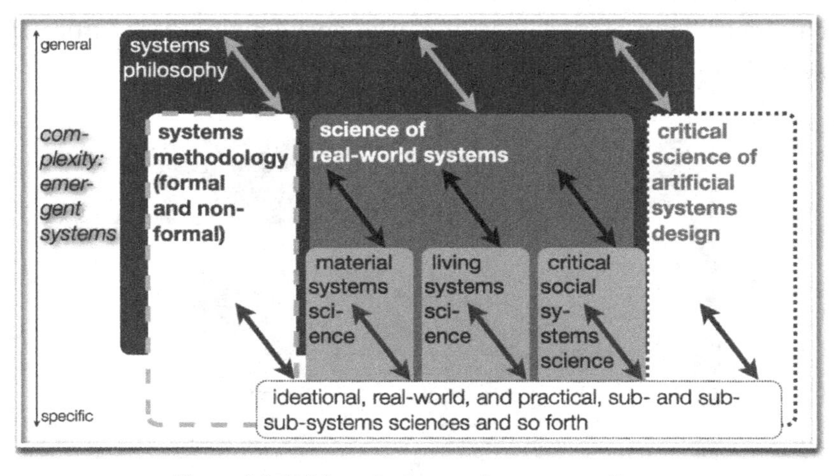

Figure 1.2. Edifice of sciences. Systems paradigm.

In a single stroke, connectedness between all inhabitants of the edifice is unveiled because emergentist systemism cuts across all habitats. A transdisciplinary approach is immediately visible. No discipline remains confined within its boundaries. Former boundaries appear semipermeable and transgressions from one scientific endeavour to another can be mediated by jumping forth and back over shared levels of scientific knowledge (see border lines and arrows in Figure 1.2). Jumping from one specific level to a more general level (that functions as a meta-level) helps compare and adjust both the initial and the meta-level. It allows the initial level to instigate knowledge adaptations on the meta-level and, equally, promotes adoption of knowledge on re-entry from the meta-level. Furthermore, the meta-level functions as a bridge for hopscotching up the ladder to even higher meta-levels or down to other specific levels or even more specific levels than one's own. This permits the insight that their knowledge is merely another specification of the knowledge of the shared meta-level. This makes it easier to enter those levels and simplifies deliberate adaptations there or considering adoptions of their knowledge when re-entering the initial levels. All these advances are mediated by the respective meta-levels. Transdisciplinarity emerges through continuous reciprocal tuning between all participating disciplines via the specification hierarchy formed by the successive levels and meta-levels.

The same holds true for the next perspective, namely that of the informational systems that build upon the emergent systems perspective (Figure 1.3). Systems philosophy can be specified to a systemic philosophy of information, systems methodology to a systemic information methodology, the science of real-world systems to a science of information in real-world systems (material, living or – in a critical perspective – social information), and critical science of artificial systems design to a critical science of information design in artificial systems. Critical Information Society Theory and Critical Techno-social systems Design Theory – whose tenets are developed in this book – are part of sciences that are subsets on the more specific end of the edifice, enjoying all the options of transdisciplinarity (Figure 1.3, bottom).

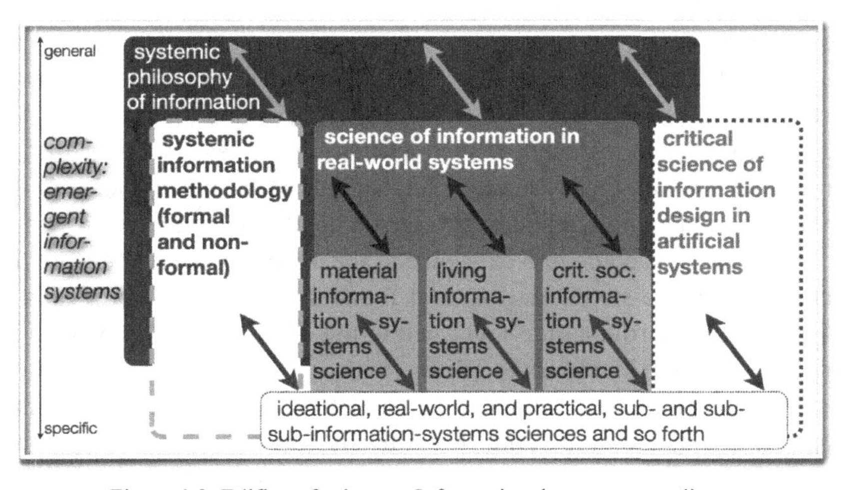

Figure 1.3. Edifice of sciences. Informational systems paradigm.

1.1 The Logic of the Third

The overview presented in this chapter is designed to recapitulate the framework principles of the *Emergent Information* book. This is the necessary starting point for the new principles presented in the subsequent chapters. Importantly, all the principles of the framework, whether articulated in the first book or formulated in present one, follow one and the same logic. This logic is the Logic of the Third.

The substantive "Third" is used here as a neuter noun. It designates neither a single person nor a plurality of persons nor a god. Instead, it designates a meta-level. A meta-level is a structure that emerges from a process that becomes, in turn, an underlying level of the meta-level. This process is driven by an interaction of entities. These entities that were a necessary basis for the establishment of the meta-level continue to form together, as a level, the necessary basis for the further existence and development of that meta-level. The meta-level therefore remains dependent on the level below. Importantly, however, after emergence, this meta-level begins to exert a dominance over the lower-lying one in that it remakes the quality of, and brings about changes to, the entities that constitute that level. These changes are such that they help better sustain the new meta-level.

Why is the meta-level called a Third? Any entity that ultimately takes part in giving emergence to a meta-level can be called a First when it is considered a single for itself. If such a First joins at least another First, it forms together with those Firsts a level that only if and when it leads to the emergence of a meta-level does earn the name of a Second. For only after the meta-level has come into existence as the Third do the Second and the Firsts prove to be steps towards, and foundations of, the Third. This is reflected in their role as the level below the Third and in their roles as constituents of the level for the Third.

Levels can be subjective, ideational in nature, but need not be so. They can also be objective, material. Meta-levels can resemble the metanoic ones that denote a subjective, ideational shift in the quality of thinking, e.g., a revolutionary paradigm shift in sciences. Equally, meta-levels can also resemble the metamorphic ones that refer to an objective, material shift in the quality of physical, biotic or social evolution, e.g., a revolutionary leap in the course of anthroposociogenesis. Furthermore, levels of the metanoic type can translate into those of the metamorphic type. This is the case for example when a revolution in science brings about a revolution in social relations with nature. Moreover, levels of metamorphosis can translate into levels of metanoia, e.g., when a revolution in human practice brings about a revolution in human thought. Finally, these two ways of comprehending levels can change places. Levels and shifts in sciences can be looked upon as objective, material ones inherent in the inquired social evolution. Equally, sciences can consider levels and shifts in natural evolution to be (proto- and quasi-) subjective ones, (proto- and quasi-) semiosic ones, when regarding the perspective of the affected entities and events.

This implies that the term "logic" in Logic of the Third should not signify laws of inferences that human thought or scientific reasoning must obey. Rather, first and foremost, in the sense of Joseph E. Brenner's Logic in Reality [2008; Brenner and Igamberdiev 2021], it signifies regularities of level formation that must be respected throughout the real world. The human mind or scientific research are not exemptions from the real-world formation of levels. It is for that reason that, from a history and philosophy of science point of view, the Logic of the Third is inherent to the new paradigm in the three different aspects developed in the subsections below.

1.1.1 *The Logic of the Third in the Designing Relation*

In order to illustrate the Logic of the Third, it is helpful to cite an idea of theoretical biologist Robert Rosen [1991; 2012, 85-211]. According to his "Modeling Relation", a natural system's internal causality can be modelled by a formal system's inferential structure such that there is an encoding activity from the natural to the formal system and a decoding one in reverse order. Encoding and decoding would be "unentailed", that is, not transferring a logically compelling conclusion [1991, 61-62]. Entailment would apply only to causality for itself and the inferential structure for itself.

The corresponding figure Rosen drew [1991, 60; 2012, 72] shows the natural system on the left and the formal one on the right, the encoding activity as an arrow from the left to the right on the lower side, and the decoding one as an arrow from the right back to the left on the upper side. The result is a flow that runs counterclockwise starting from the natural system to the formal one and, via inferences in the formal one, returns to the natural one again as its destination.

That figure helps draw a new one that better fits the requirements of the new science paradigm. In that paradigm, the idea of a subject-object dichotomy has given way to a dialectic comprehension of the subject-object relation. Far from abandoning the distinction of subject and object, the new comprehension concedes that the subject is part of the objective world and that the world of objects is part of a subjective approach. This does justice to the ineluctable interdependence of subject and object, without denying the fundamental asymmetry that forestalls a voluntary changing of their places.

Because the subject is intrinsically interwoven with the objective world, she has a subjective take on the world: the objects have a meaning for her. She has subjective needs that are given to her objectively as well as desires that emerge objectively. Accordingly, she is confronted by objects that do not afford her needs and desires and must cope with those challenges. Mechanistic solutions are rare. They fit only those challenges that are not complex. Complex challenges, however, require a lift up to a higher level. The task is to find such a higher level, which is simply equivalent to finding a Third. Science is a way – and it is the best way – to search for possible Thirds.

For that very reason, Rosen's Modeling Relation is extended here to a Designing Relation and changed according to the subject-object dialectic (Figure 1.4). The natural system is replaced by objects that might be adapted and therefore scientifically investigated; the encoding activity is replaced by an ideational activity of the subject; the formal system is replaced by the scientific knowledge of the subject; and – this is the most important and innovative consequence – the decoding activity is replaced by a materialisation of the scientific knowledge as a meta-level on top of the object level that is to be adapted.

Science is set here to intervene in processes and structures that matter to society. The intervention itself is a social process that strives to form a structure in processes where no structure previously existed, or to transform one structure into another one. The impact of science is seen as implanting subjective ideas as Thirds into the realm of the objects such that these ideas materialise there.

This is how, praxiologically, the Logic of the Third can be found in the new paradigm.

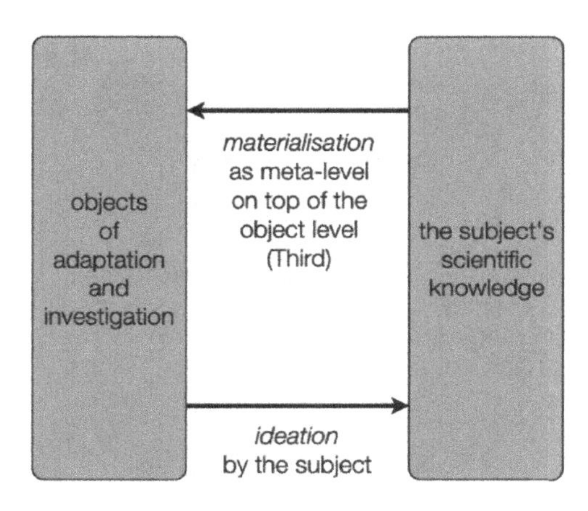

Figure 1.4. The Designing Relation. The praxiological Third of the new paradigm.

1.1.2 *The Logic of the Third in the Modelling Relation*

The praxiological Third is not the only Third that characterises the new approach to science. In the subject-object dialectic the object plays a part of its own that is covered by the new Modelling Relation (Figure 1.5). The objects of adaptation and investigation for the subject turn out to be real-world objects that themselves consist of events and entities, of processes and structures, in particular of processes that give rise to structures that, in turn, channel those processes. Hence, they are recognisable as containing levels and meta-levels. There are numerous meta-levels – objective Thirds – of interest to the subject.

Science is set here to understand objective Thirds because these Thirds decisively influence how the objects work. And the better these Thirds are understood, the better a subjective Third can be ideationed for transferral onto the objects to become a new material Third, which in turn adapts those objects.

From this perspective, Rosen's causal entailments need to be replaced too by relations of emergence and dominance according to a less-than-strict determination rather than relations of necessity only.

This is where the ontological Third of the new paradigm proves the Logic of the Third.

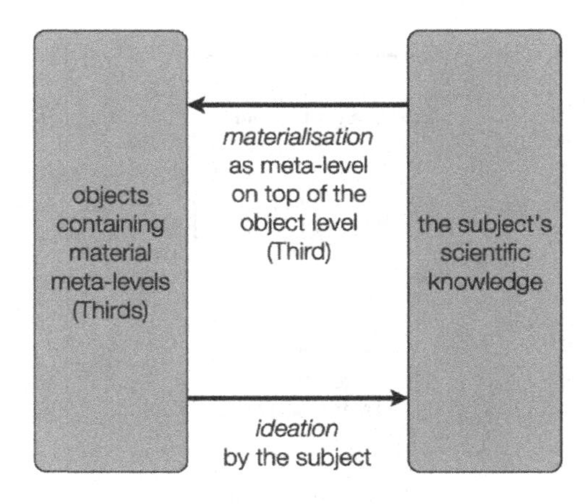

Figure 1.5. The Modelling Relation. The ontological Third of the new paradigm.

1.1.3 *The Logic of the Third in the Framing Relation*

Last but not least, there is also an epistemological Third (Figure 1.6). This epistemological Third consists of how the subject works when generating scientific knowledge. Scientific knowledge must run through different kinds of ideational quality before it reaches the loop in which it is materialised. These qualities are ordered in levels. Thus, meta-levels are inherent. These meta-levels do not, as such, reflect, i.e., mirror, material meta-levels contained in the objects. The process of generating scientific knowledge contains subjective meta-levels that are specific solely to this very process and differ in kind from objective meta-levels. Nonetheless, and even inherently, by passing through its meta-levels that do not correspond to the objective meta-levels, this process can achieve a reflexion of the latter ones.

Science is set here to interrogate objects containing meta-levels that matter to society. The interrogation itself is an ideational process that is not merely a one-to-one, mechanical reflection but a complex, creative human/societal reflexion with leaps in quality. It passes through a procedure designed to yield higher information structures based upon lower levels.

The introduction of such meta-levels also replaces Rosen's inferential structure. It is replaced by a hierarchy of three levels: the topmost level accommodates the aims of scientific knowledge application, the intermediate level accommodates the scope of scientific knowledge mapping, and the bottom level accommodates the tools of scientific knowledge generation. The aims level directs the application of scientific knowledge towards a technology that is appropriate to work as a meta-level that is mindfully inserted into the object. That enables the object to better suit society's needs and desires and, at the same time, to better do justice to the *eigensinn* of the objective processes and structures with which humans interrelate. The scope level directs the mapping of scientific knowledge towards a theory that is, after Karl R. Popper, falsifiable but allows for appropriate technologies on the next level. The tools level directs the generation of scientific knowledge towards a methodology that is multi-faceted to avoid one-sidedness when generalising the empirical findings onto the theories level.

All levels are connected through bottom-up and top-down cycles that are, in Rosen's sense, unentailed so as to comply with relations of emergence and dominance devoid of strict determinism. The generation of scientific information runs over cycles from below and from above that connect the aims, scope and tools levels. The lower levels are the necessary preconditions for the upper ones. This lets content emerge from below, while at the same time enabling the upper levels to shape the lower ones so that their content is dominated from above, too. This is visualised in the figure by arrows: those from below signify emergence and those from above dominance, albeit in a not-entailing way.

The introduction of the term Framing Relation here is to indicate that the Logic of the Third comes to the fore on the subjective side of the subject-object-dialectic of science.

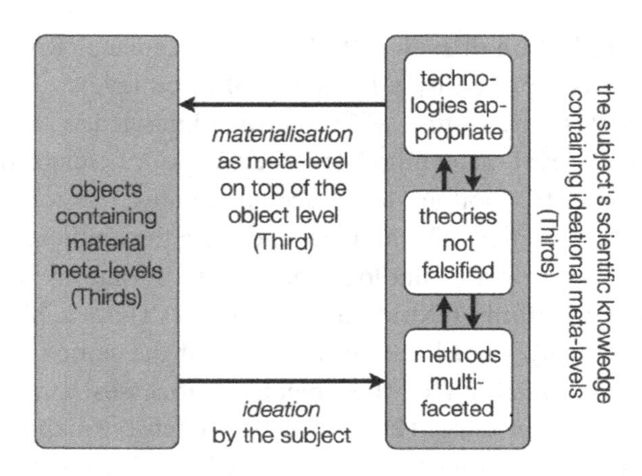

Figure 1.6. The Framing Relation. The epistemological Third of the new paradigm.

The Logic of the Third takes centre stage in the philosophy, in the systems thinking and in the study of information, all of which form the core of the new paradigm introduced in the book *Emergent Information*. Viewed from a history of science perspective, the paradigm shift promoted here is a shift from the often-called "Cartesian/Newtonian" normal science to emergentism as the praxio-onto-epistemological turn in philosophy; a shift to emergentist systemism, represented by the evolutionary-systems-science turn in systems thinking; and a shift to emergentist systemist informationism, that is, the science-of-information turn in the study of information (with the Unified Theory of Information – UTI – as a building block). This shift is expressed by principles that guide the production of the content in the comprehension of the world, of systems and of information. The next sections review these three steps of the new paradigm in the light of the Logic of the Third. They detail the Logic of the Third with respect to the aims, scope and tools levels on which designing, modelling and framing take place.

1.2 Emergentism: Unity-through-Diversity as Third

The philosopher of science Mario Bunge defined emergence as "qualitative novelty", more precisely, "its occurrence in the course of some process" and as "property of a complex object [...] if neither of the constituents or precursors of the object possesses it" [2003, 17; see also 2012, 185].

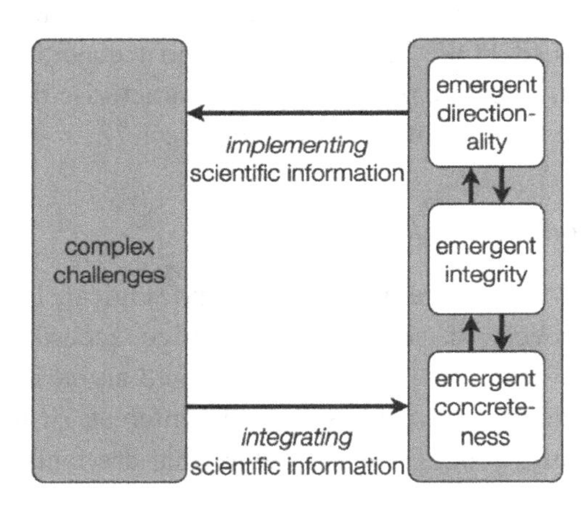

Figure 1.7. Emergentism. The praxio-onto-epistemological content of the new paradigm.

In Bunge's vein,[a] not every ism is a dogmatic one [2012, 30]. Accordingly, emergentism is coined here to characterise a programmatic ism that includes a particular *weltanschauung*, that is, a world view that is not value-free because it gives directives on how to tackle problems (*weltanschauung* is a German term used by Mark Davidson [1983] to sum up Ludwig von Bertalanffy's General System Theory and which is subsumed here under praxiology). Emergentism therefore describes a particular conception of the world and its working (ontology) and a particular way of thinking that generates knowledge about the world

[a] Bunge himself was inclined to use the adjective as in "emergentist materialism" to signify his position in the field of the mind-body problem or to use the noun synonymous with systemism [2003, 38-39]. Here, however, emergentism will be considered to be independent of systemism.

(epistemology). Emergentism, then, is a *weltanschauung*, a world conception and a methodology of worldly knowledge creation that revolve around the occurrence and the properties of qualitative novelty. These three tenets encapsulate each other in line with the aims, scope and tools levels of Praxio-Onto-Epistemology (POE). POE is the highest level-of-abstraction endeavour to integrate scientific information about complex challenges in order to implement scientific information to intelligently cope with complexity. Directionality, integrity and concreteness are the emergent features of POE. They form the most general foundations of the new paradigm. They are the Thirds that characterise the designing, the modelling and the framing of the world (Figure 1.7).

1.2.1 *Emergent directionality*

Directionality belongs to the uppermost, praxiological, level of the three ideational meta-levels of scientific knowledge according to the new paradigm. It is twofold. There is an extrinsic and an intrinsic direction of facts. The extrinsic direction stems from the interests or the needs of the human observer of a process or a structure that are facts. This direction consists in how the facts enable satisfying those interests or needs. The intrinsic direction stems from the interests or the needs of human or other agents[b] that are involved inside the process or the structure that are facts. This direction consists in how the facts enable satisfying these interests or these needs, whether or not observed by an outside human observer. For either directionality, the same considerations of how the factual and the directional are put together are valid – the consideration of the descriptive and the normative, or of Is and Ought, as generically labelled in philosophy (Table 1.1).

---

[b] Agents are things that are capable of agency. As a rule, agency is associated with autonomy, spontaneity, contingency, as these features are not subject to strict determinacy. Agents belong to different realms of the real world such as the physicochemical, the biotic, the human.

Table 1.1. The consideration of Is and Ought in emergentist praxiology. The factuals and the directional.

		is	**ought**
con-flation	**reduction: naturalistic fallacy**	**the factuals:** sufficient conditions for the directional	**the directional:** resultant of the factuals
	projection: animist fallacy	**the factuals:** resultants of the directional	**the directional:** sufficient condition for the factuals
disconnection: nihilistic fallacy		**factuals**	**directional**
		disparate takes	
combination: emergentist praxiology		**the factuals:** necessary condition for a directional	**the directional:** an emergent from the factuals

Relations can be misunderstood through conflations and disconnects (see also Hofkirchner [2013a, 39-46]). A conflation occurs if distinct relata are viewed as equal. Such a conflation can involve reduction or projection. In the case of Is and Ought, the reduction would commit a naturalistic fallacy: the directional would – as a resultant of the factuals – be reduced to the factuals – as a sufficient condition for the directional. The projection would commit an animist fallacy: the directional would – as a sufficient condition for the factuals – be projected onto the factuals – as a resultant of the directional.

A disconnect occurs if connected relata are viewed as unconnected. Accordingly, a nihilist fallacy would declare the factuals and the directional to be disparate takes.

The understanding that does justice to both the identity and the difference of the relata is an emergentist combination Emergentist praxiology assumes that factuals are a necessary but not sufficient condition for a directional that, in turn, is not a resultant but an emergent from the factuals. The Ought goes irreducibly beyond the Is, but remains grounded in the Is.

The Ought possesses the property of a Third, an emergent meta-level, compared with the manifold factuals of the Is – directionality for design.

1.2.2 *Emergent integrity*

Integrity belongs to the intermediate, ontological level. It focuses on the observed. Any process or structure is a becoming or being as an integral, that is, an integration of what is differentiated, a unitary made up of the diversified. The question is: how is being produced by becoming, how is being reproduced and what is the nature of non-being (Table 1.2) in order to give direction to the facts?

Table 1.2. The consideration of non-being and being in emergentist ontology. The causative and the caused.

		non-being	**being**
con-flation	**reduction: mechanistic fallacy**	**the efficient, or material, causative:** sufficient conditions for the modular	**the modular:** resultant of the efficient, or material, causative
	projection: teleological fallacy	**the final, or formal, causative:** resultants of the teleological	**the teleological:** sufficient condition for the final, or formal, causative
disconnection: existentialist fallacy		**nothing causative**	**something else**
		independent existents	
combination: emergentist ontology		**the present potential:** necessary conditions for a future integral	**the actual integral:** an emergent from the past potential

The reductive conflation is the fallacy of mechanicism in the case of less-than-strict determinacy (see Hofkirchner [2013a, 86-95]). Efficient, or material, causes are deemed sufficient for the transition from the non-being to a modular (but not an integral) being. The projective conflation is the fallacy of teleology, again in the case of less-than-strict determinacy. The telos is endowed with sufficiency for resulting in final, or formal, causes in the state of non-being.

The disconnect can be called an existentialist fallacy: non-being and being are viewed as nothing and something, respectively, in particular as nothing causative and something else, regardless of whether modular or integral, in which case no transition is thinkable.

Only the emergentist combination can provide a rational account: rather than being nothing, non-being is a present potential as a necessary condition for a future integral, and the actual integral is an emergent from the past potential. The causative is anchored in the potential, and the caused emerges. That is the posit of emergentist ontology.

The integral being is envisioned as possessing a Third – integrity that needs to be modelled in an integrative manner. This stands in contrast to all the differentiations at the past non-being state of precursors or the actual state of constituents.

1.2.3 *Emergent concreteness*

Concreteness is resident on the lowermost, epistemological level. It is focused on the observer and asks how he or she can bring together appearance and essence of becoming and being (Table 1.3).

Table 1.3. The consideration of the apparent and the essential in emergentist epistemology. The abstract and the concrete.

		apparent	essential
con-flation	**reduction: empiricist fallacy**	**the common descriptive:** sufficient condition for the inductive general	**the inductive general:** resultant of the common descriptive
	projection: hypostatic fallacy	**the particular:** resultant of the universal	**the universal:** sufficient condition for the particular
disconnection: agnostic fallacy		**the given**	**the imaginative**
		incommensurable knowledge	
combination: emergentist epistemology		**the abstract:** necessary condition for a concrete	**the concrete:** an emergent from the abstract

The reductive conflation is the empiricist fallacy. It underestimates theory and overestimates complete induction. Such induction is a deductive inference (and as such compelling) without yielding any new insight that goes beyond the actual description of what certain phenomena seem to have in common. The induced generalisation is a resultant of the description as a sufficient condition. The common description is an abstract and so is the inductive general. The projective conflation is the hypostatic fallacy. The universal is projected onto the particular so as to allow another compelling deductive inference, this time from the universal as a sufficient condition to the particular as a resultant. The universal is concrete, as is the particular.

The disconnect is a withdrawal from any compelling inference – the agnostic fallacy. The apparent given is treated as if independent from any essential imagination. The abstract and concrete are not mediated.

The epistemological combination recognises the interdependence of appearance and essence in an emergentist manner. Both must be reconstructed in ideation by "a combination of analytical and synthetic tools: here, analysis decomposes the real, concrete something as it can be observed (in an undifferentiated way), and synthesis recomposes the concrete from the differentiated abstract items in the way they are composed in reality" [Hofkirchner 2013a, 132]. Accordingly, the concrete is an emergent from the abstract that is, in turn, the necessary condition for the concrete. Concretisation is an ascendence from the abstract to the concrete. At each step, something abstract is added so as to form, without deduction, an ever-new concrete.

Concreteness is the Third that gives the frame its function – a guide through the seeming disorder of abstractions, helping to knit them together.

1.2.4 *The Principle of Unity-Through-Diversity*

Directionality, integrity and concreteness together build unity-through-diversity [Hofkirchner 2013a, 46]. Conflations presume unity without diversity. Disconnects presume diversity without unity. But the combinations posit unity through diversity such that diversity is a necessary condition for unity and that unity is an emergent from diversity. No unity is possible without diversity because only diversity can be united. Since unity is emergent, however, it is contingent and not a must. Unity can emerge but need not do so and, when it does, it can do so at varying times in varying ways and degrees.

Unity-through-diversity is a subjective/objective Third. It is subjective as well as objective in the case of directionality. This is because giving directions for design has two aspects. On the one hand, a subjective take of an outside observer on objective processes or structures might afford the satisfaction of the observer's interests or needs. On the other hand, these very processes or structures already display directions given by the object itself as agent, be they human or other agents.

These directions aim at reaching states of unity-through-diversity in the processes or structures.

In the case of integrity, unity-through-diversity is primarily an objective Third. It is objective because every becoming is inclined to integrate processes or structures into a whole, and every being is an integrated whole inclined to continually preserve its integration. Note that stagnation, declines, and disintegration can occur. Unity-through-diversity is an objective integral – the identity of the differentiations: here, a convergence to which processes tend though divergence can occur, or the emergence of a new level though submergence to precedent older levels can occur [Bunge 2003, 17]. Secondarily, unity-through-diversity is a subjective Third. It is subjective because modelling the integration of an object is carried out by the subject based on integrative thinking [Hofkirchner 2013a, 46]. Unity-through-diversity is a subjective integral – an ideal to which materialised integrals can be compared.

In the case of concreteness, unity-through-diversity is primarily a subjective Third. It is subjective because the ascendence from the abstract to the concrete is an ideational method of the subject to enable the concrete to emerge in thinking. Secondarily, however, it is objective because this kind of framing through ascending from the abstract to the concrete will reconstruct in the idea how the object has been constructed in reality – the concretisation is the subjective framing of the objective concrescence, the objective growing together of processes or structures.

In conclusion, unity-through-diversity, both subjectively and objectively, turns out to be the praxiological directional, the ontological integral, and the epistemological concrete in emergentism. This is the new philosophy of the new paradigm.

After this more detailed discussion of emergentism, the following definition can be given:

Emergentism. *Emergentism is that* weltanschauung, *that conception of the world and that way of creating knowledge of the world that applies the Principle of Unity-Through-Diversity.*

The **Principle of Unity-Through-Diversity** states: there is, subjectively (as a matter of human agents' take) and objectively (as a matter of fact to be assumed for human agents regardless of whether or not they actually observe it),

(1) directionality in human agents' knowledge about worldly processes and structures, becoming and being, as well as in the very worldly processes and structures, becoming and being, themselves, originating from the agency of human or other worldly agents;

(2) an inclination towards integrity of worldly processes and structures, becoming and being, and of human agents' knowledge about them;

(3) concreteness of creating human agents' knowledge about, and of, worldly processes and structures, becoming and being;

such that the directional, the integral, and the concrete – nesting one another in succession – are meta-level emergents, that is, Thirds. These Thirds represent wholes, actualised in ideations or materialisations by interaction networks as plural Seconds, in which participating human or other worldly agents represent single Firsts. Any whole is a unity emerging from the diversity of its precursors or constituents.

1.3 Systemism: Self-Organisation as Third

Systemism is a term defined by Bunge [2012, 189] as follows:

> Ontology: Everything is either a system or a component of some system. Epistemology: Every piece of knowledge is or ought to become a member of a conceptual system, such as a theory. Axiology: Every value is or ought to become a component of a system of interrelated values.

For him, systemism is synonymous with emergentism [2003, 38-39]:

> Systemism, or emergentism, is seen to subsume four general but one-sided approaches: 1. *Holism* [...] 2. *Individualism* [...] 3. *Environmentalism* [...] 4. *Structuralism* [...] Each of these four views holds a grain of truth. In putting them together, systemism (or emergentism) helps avoid four common fallacies.

In the context of this book, systemism shall be used explicitly with an emergentist meaning (systemism would not make sense without emergent systems, but emergentism is not per se systemist). "Emergentist systemism" is the term of choice. It was introduced by Poe Yu-ze Wan [2011] in the context of social theory, after it had been used in the field of social work in Switzerland and Austria [Obrecht 2005] to characterise Bunge's approach as well as the Salzburg approach of Hofkirchner and collaborators developed from 2004 to 2010 [Hofkirchner et al. 2007]. It is used here in a somewhat broader sense than intended by Wan. Systemism is the next step after emergentism to characterise the paradigm shift.

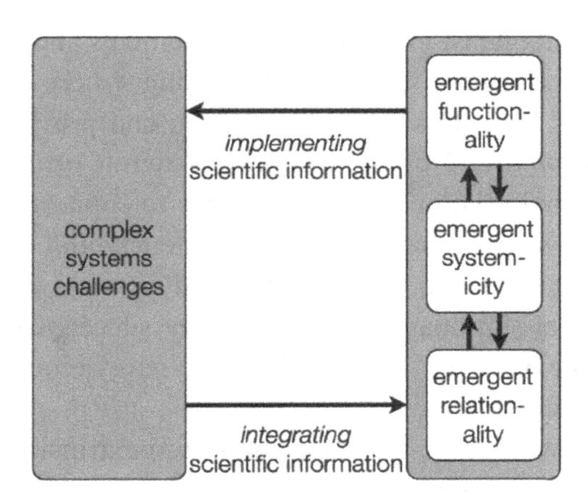

Figure 1.8. Emergentist systemism. The evolutionary-systems content of the new paradigm.

Systemism studies the systemics of the world. It is understood as a *weltanschauung*, a world conception and a way of creating world knowledge. According to that *weltanschauung*, humans typically strive to design real-world systems to harness their synergy effects for the betterment of social life. They also model their systemic functioning as a nested self-organisation of systems that become, or are, elements of other systems, and they frame the systems' working based on an understanding of their organisational relationality. Synergetic functionality, a dynamic systemicity and relationality are the three Thirds that are emergent features argued by the Evolutionary Systems Theory (EST), built upon directionality, integrity and concreteness. Together they form the Principle of Self-Organisation.

1.3.1 *Emergent functionality*

A core tenet of understanding systemics is that there is no complex system without objective function. Objective functions come about with synergy. Synergy occurs whenever a common effort of agents is rewarded by an effect unachievable in isolation. Agents can be of any possible type. They organise themselves into relations that enable the advent of synergetic effects. Those relations consolidate and a system arises, i.e., is formed and then maintained, if agents, initially as proto-elements and subsequently as elements, benefit from the system by providing synergy. The system is upheld as long as they do so. The production and provision of synergy effects is the *raison d'être* of any complex system [Corning 1983; 2003]. If the organisational relations no longer serve to support the elements in producing and providing synergy, then the system would break down. The currently existing physical, living and social systems are outcomes of natural or social evolutionary processes. They succeeded in being most enduring by virtue of their ability to adapt to synergy requirements. Those that were less successful vanished.

In analogy with the differentiation of extrinsic and intrinsic directionality, the phenomonen of synergy can be treated in either way. Humans and society can take advantage of the synergy production of natural or social systems in that they intervene in the synergy production to further it. That is a design action changing the natural or social system such that the synergetic procedure is embedded in an overarching social system. Sometimes the intervention does not propagate synergy but, willingly or not, impedes, halts and even destroys its proliferation.

Regardless, synergy confers directionality to systems. Synergy is the overall objective function of a complex system.

The processual and the functional are the specifications of Is and Ought in the systemist praxiology (Table 1.4).

Table 1.4. The consideration of Is and Ought in systemist praxiology. The processual and the objective functional.

		is	**ought**
con-flation	**reduction: naturalistic fallacy**	**the processual:** sufficient condition for the objective functional	**the objective functional:** resultant of the processual
	projection: animist fallacy	**the processual:** resultant of the objective function	**the objective functional:** sufficient condition for the processual
disconnection: nihilistic fallacy		**processual**	**objective functional**
		disparate takes	
combination: systemist praxiology		**the processual:** necessary condition for a synergetic function	**the synergetic functional:** an emergent from the processual

The naturalistic and the animist fallacy conflate dynamics and objective function through reduction and projection. This parallels their mistake with regard to the praxiology of emergence: either the objective function of the system that gives a direction is reduced to, or is projected onto, the dynamics of the system. The nihilistic fallacy then disconnects them from each other as if discordant grasps, whereas the systemist praxiology relates them through emergence. That praxiology considers the systemic dynamics as a necessary condition from which the fundamental systemic objective function of creating synergy can emerge.

Comparably with the Ought in emergentist praxiology, the systems' striving for synergy as an Ought can be considered a Third – it is a meta-level that adds a new effect to the interaction of systems at a lower level.

1.3.2 *Emergent systemicity*

The synergistic answer to the question of how the objective function of a system and the system's dynamics are related has implications for modelling such dynamics. Here, the term systemics denotes the overall dynamics of complex systems, in particular with respect to the ontological question of how to reflect upon non-being and being of systems, i.e., how systems become and how they manage to maintain their being.

Of course, systems are wholes that underlie push-and-pull causalities of drivenness and end-directedness as well as materiality and formative power according to the four causes after Aristotle [Hofkirchner 2013a, 89-91]. Since modern science before the paradigm shift has rejected further utilising three if not all four of these causes, this science approach represents fallacious moves or no move at all when it comes to systems. Initial and boundary conditions, other factors and the possibility space are modelled as points of reference to better understand the dynamics of systems that display trajectories from one system state to another (Table 1.5).

Table 1.5. The consideration of non-being and being in systemist ontology. The conditional and the transformational.

		non-being	**being**
con-flation	**reduction: mechanistic fallacy**	**the initial and boundary conditional:** sufficient condition for the transformational	**the transformational:** resultant of the initial and boundary conditional
	projection: teleological fallacy	**the initial and boundary conditional:** resultant of the transformational	**the transformational:** sufficient condition for the initial and boundary conditional
disconnection: existentialist fallacy		**conditional**	**transformational**
		independent existents	
combination: systemist ontology		**the possibility-space conditional:** necessary condition for a self-organised transformational	**the self-organised transformational:** an emergent from the possibility-space conditional

In the mechanistic interpretation that is still mainstream in materialistic science, state changes are reduced to initial and boundary conditions that represent cause-effect-relationships between system variables. The latter can be mathematically formalised to compute next system states. Within certain limits, such models can simulate the development of real-world systems with a high degree of accuracy and yield convincing forecasts.

For those models, however, the real-world evolution of systems is out of reach because they cannot predict – or explain – the advent of new properties.

The opposite of materialistic science, the teleological belief, does the opposite yet also fails to explain – or predict – the advent of new properties. This is because idealism projects the next system states onto the initial and boundary conditions in which the former are not yet realised.

Another variety of idealism refuses to identify factors that are relevant for state changes and thus cannot explain systemics either. It disjoins the being of systems from their becoming by cutting the non-being back to nothing.

This non-being is, according to systemist ontology, identified as the space of possibilities of systems-to-be that is given with any realisation of systems at any level. This space contains the necessary conditions for the advent of a new system on a higher level or of a new state of the system at the same level of evolution. This is because something can be actualised only if existent as potential – from nothing, nothing can come. What is actual, is, in the last resort, contingent, since it is emergent. A spiral upwards can be set in motion – driven by self-organisation loops. Self-organisation signifies the dynamics of complex systems, the systemics of the becoming of a complex system and of the being of a complex system (Hofkirchner 2019, 4):

> There is a bottom-up process in which the interactions of elements (or proto-elements in the case of emerging systems) cause the emergence of relations of organization that become solidified on a higher level. Equally, there is a top-down process in which these solidified organizational relations exert a causal power on the activities of the elements. Thus, after the forming of a system, that very system maintains itself such that— through downward causation—its organizing relations make its elements produce the system itself anew. And the elements—through upward causation letting organizational relations emerge—maintain themselves by making the system organize relations for the production of the elements.

Self-organisation means:

> The system (the self) refers to itself in that it lets its organizational relations refer to its elements (many selves). These, in turn, refer via the organizational relations to the system. [Hofkirchner 2019, 4]

Systems are therefore true systems only if they are emergent systems, i.e., systems whose dynamics include emergence, a qualitative novelty, when being formed as a meta- or suprasystem and when changing its states in the course of evolution.

Such an emergent system is defined as "*a collection of (1) elements E that interact such that (2) relations R emerge that – because of providing synergistic effects – dominate their interaction in (3) a dynamics D.* This yields a distinction between micro-level (E) and macro-level (R) and a process (D) that links both levels in a feedback loop" [Hofkirchner 2013a, 105].

The self-organisation loop is key to any understanding of emergent systemics. It embodies the Third that makes agents re-appear as systems, and it assembles the integrity of systems.

Self-organisation underwent an evolutionary process. In brief, in the systems world that humanity is currently confronted with, material self-organisation has unfolded into self-organisation of non-living and living material systems and the latter into self-organisation of pre-human and human social living material systems. They differ according to the end-directedness and the formative power characteristic of those system types [Hofkirchner 2013a, 90-91].

1.3.3 *Emergent relationality*

The concept of systemics including self-organisation loops presupposes the concept of organisational relations. Bertalanffy's ground-breaking idea reconciled the mechanicism-vitalism dispute in biology and paved the way for the General System Theory beyond organismic systems of the biotic realm. It recognises that the organisation of what traditionally has been called substances is what makes the difference between life and non-life and not the substances as such. Bertalanffy theorised "organising relations" that would unite such substances into systems [Bertalanffy 1932, 81, 98].

> The properties and behaviour of a system cannot be explained by a "summation" of the properties and behaviour of the parts when investigated independently from the system. In a system, the behaviour of the parts depends on how they are related and cannot be predicted in advance. Although a sum can be put together gradually, a

system is given with its parts and their relations at one blow. [Hofkirchner 2019, 264]

Without conceiving the relations, no system can be conceived, as Bertalanffy formulated later with the focus on organisms as types of systems – but, of course, this is generalisable for any kind of system [1950, 134-135]:

> As opposed to the analytical, summative and machine theoretical viewpoints, organismic conceptions have evolved in all branches of modern biology which assert the necessity of investigating not only parts but also relations of organisation resulting from a dynamic interaction and manifesting themselves by the difference in behaviour of parts in isolation and in the whole organism.

From this quote, it becomes clear that Bertalanffy distinguished between a dynamic interaction of the parts, on the one hand, and relations of organisation, on the other. Thus, the interactional and the relational are those categories whose connection needs to be discussed when specifying appearance and essence in systemist epistemology (Table 1.6).

Table 1.6. The consideration of the apparent and the essential in systemist epistemology. The interactional and the relational.

		apparent	**essential**
con-flation	**reduction: empiricist fallacy**	**the interactional:** sufficient condition for the relational	**the relational:** resultant of the interactional
	projection: hypostatic fallacy	**the interactional:** resultant of the relational	**the relational:** sufficient condition for the interactional
disconnection: agnostic fallacy		**interactional**	**relational**
		incommensurable knowledge	
combination: systemist epistemology		**the interactional:** necessary condition for an organisational relational	**the organisational relational:** an emergent from the interactional

The empiricist fallacy holds interaction – instantiating the common descriptive backed by philosophy of the old paradigm – as a sufficient condition for the relations. This instantiates the inductive general as the conclusive consequence. The hypostatic fallacy executes the conflation of appearance and essence the other way round. The relations – here instantiating the universal in the old philosophy – are said to be the sufficient condition from which the interaction (here instantiating the particular) can be concluded.

For the fallacy that is used to dispense procurable insights, interaction and relations represent incommensurable knowledge.

Bertalanffy's emphasis on organisational relations is a matter of theory. "Science is not a mere accumulation and catalogue of facts. It is a conceptual order we bring into facts" [Bertalanffy 1953, 238].

> The relations of organization in a system – though endowed with the properties of having been caused by a dynamic interaction of the elements of the system and of exerting causative power on the behaviour of each element in which the relations manifest themselves – are not observable, in contrast to the dynamic interaction or the behaviour of the elements. These relations need to be construed theoretically, and they are necessary in promoting understanding and explaining the empirical data of interaction and behaviour – data that would be senseless without interpretation in the light of organizational relations. [Hofkirchner 2019, 3]

The relational is an emergent from the interactional, and the interactional plays the role of a necessary condition. The relational cannot be concluded from empirical data in a deductive manner. It needs to be theorised. Once certain organisational relations have been hypothesised, the construct can be corroborated by facts.

Relationality is a Third. It is an elaboration of concreteness. Emergent systemic relations are the glue that brings the elements of a complex system together.

1.3.4 *The Principle of Self-Organisation*

Functionality as systemist elaboration of directionality, systemicity as systemist elaboration of integrity, and relationality as systemist elaboration of concreteness: they revolve altogether around systems that organise themselves.

Old paradigmatic reductionism tries to boil objective functions, states and relations of systems down to epiphenomena of a mechanistically perceived dynamics. Traditional projections know only hetero-organisation. Conventional disjunctivism cannot bridge the gaps. Only the emergentist-systemist approach presents an account of true self-organisation. Systems are individual agents that are made up of other systems and can work together collectively to build up a higher system. They are selves that can organise another self.

Self-organisation is a subjective/objective Third. Its subjective and its objective side pop up at the same time when human systems take advantage of other systems, yielding synergy for themselves. In doing so, they exercise extrinsic directionality. The emergent objective function of those systems – their intrinsic directionality – thus becomes an emergent objective function of human systems.

Furthermore, the objective side of this Third shows up in emergent meta-system transitions [Metasystem Transition], which produce integrity, and in state transitions that emerge during the evolutionary development of suprasystems [Hofkirchner 2013a, 118], which reproduce integrity. This reflects the organisational activities of systems, between systems and in systems. The subjective side of systemicity appears in numerous attempts of human systems to understand the loops of self-organisation.

Finally, the subjective side of the Third of self-organisation is displayed in the search by human systems for knowledge emerging about the systemic relations that play the decisive part in understanding the dynamics of self-organisation. These relations represent the indispensable essence of the concreteness. This is so because, according to the objective side, the relations of organisation that are (re-)produced by the activities of the system's elements decide about the new production and provision of synergy.

Self-organisation strives for unity-through-diversity, exercises unity-through-diversity and construes unity-through-diversity. Self-organisation is the praxiologically functional, the ontologically systemic and the epistemologically relational of emergentist systemism – the new system theory of the new paradigm (EST).

The following definition can be given:

Systemism. *Systemism is that* weltanschauung, *that conception of the world and that way of creating knowledge of the world that applies the*

Principle of Self-Organisation, based upon the emergentist Principle of Unity-Through-Diversity.

> The **Principle of Self-Organisation** states: there is subjectively (as a matter of human systems' take) and objectively (as a matter of fact to be assumed by human systems, regardless of whether or not they actually observe it),
>
> (1) a functionalisation of systems of any kind by human systems as well as by systems themselves concerning synergy;
>
> (2) a systemic functioning throughout the real-world through loops from systems of one level as proto-elements to a metasystem they build and as elements of a suprasystem they inhabit on a higher level and from there back to the lower level. This is systematically covered by thought in a hierarchy of levels to be modelled;
>
> (3) a relevance – for the comprehension of systems, of relations of organisation – that make the elements function in service of an objective function;
>
> such that the functional, the systemic, and the relational – thereby specifying the directional, the integral, and the concrete – are nested meta-level emergents, that is, Thirds. Those Thirds represent self-organised systems, actualised in ideations or materialisations by interaction networks as plural Seconds, in which participating human or other systems represent single Firsts. Any system organises itself via the organisation carried out by its proto-elements or its elements.

1.4 Informationism: Information as Third

The first step (section 1.2) was to revisit the world as an emergent unity-through-diversity architecture that is constituted by a manifold of worldly agents (POE). This was followed by identifying those agents as self-organising real-world systems that strive to produce and provide synergy effects (EST) in section 1.3. The next step is to clarify the role of information (in the wake of the Unified Theory of Information – UTI – of book one). The basic idea is to view those systems as also being capable of informational agency, depending on the class of systems to which they pertain and according to the evolution they underwent. Emergent systems generate emergent information. Thus, the generation of information is enshrined in the self-organisation of systems. This shall be referred to as informationism here. Again, this is to be perceived as programmatic ism[c] – as *weltanschauung*, world conception and a way of understanding that builds upon emergentism and emergentist systemism. Emergentist-systemist informationism is a guidance for the design of systems of whatever kind by human systems – under the condition that the systems to be designed generate specific information processes for themselves; it is a map of how information is generated; and it is a method to recognise information generation. These items are discussed under the terms valuableness, reflectivity and semioticity (Figure 1.9).

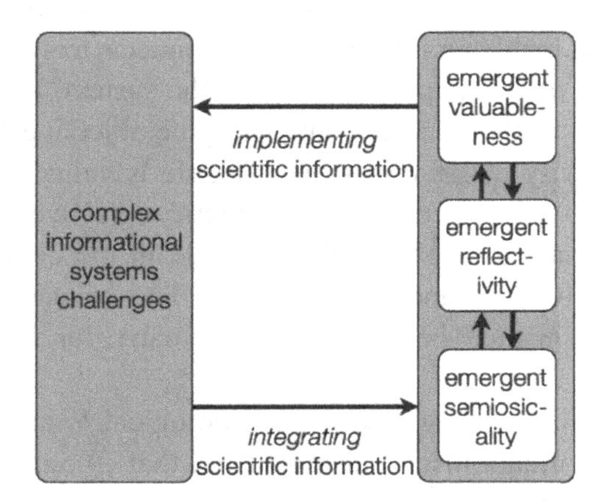

Figure 1.9. Emergentist-systemist informationism. The emergent-information content of the new paradigm.

1.4.1 *Emergent valuableness*

It is appropriate to speak of an ideational good in the context of human systems. Here, human systems signify a match of their interests and needs with the affordance of the nature of systems other than themselves to meet their interests and needs. This ideational good represents a positive evaluation of an object. It is itself subjective in that it depends on the interests and needs of the subject, and it is objective in that it reflects upon

[c] Whereas Bunge defines informationism as dogmatic ism, namely, as a hype that examines information "regardless of material things" [2012, 31].

the affordance of the object. The evaluation is made in grades, that is, it can value the object more or less good for the subject. Furthermore, it can judge an object as being irrelevant or even as more or less the opposite of good, namely bad. This goodness ideation of human systems builds upon the human take on systemic synergy functions. That take elaborates the extrinsic unity-through-diversity direction of agents.

Importantly, there is also an intrinsic equivalent of directionality. It extends over the objective functionality of synergy for any human as well as nonhuman system to a general valuableness that is a feature of any informational system. The evaluative goodness is merely the human-specific case of valuableness. Valuableness connects to synergy, whereby synergy becomes the principal valuable for the system. It is the valuable that makes the system and its elements serve the objective function of the system. Any activity that furthers the valuable is valuable as well. The valuable accrues and concerts the synergetic objective function. This makes it appropriate to speak of the valuable in the context of natural systems as well [Collier and Stingl 2020].[d] The valuable appears as a structural pattern in their behaviour. It is valuable for them to respond appropriately.

The valuable is the Ought of the praxiological Is-and-Ought in the perspective of informationism. The affordance that is cognised, that which is observed to be afforded, the cognisantly affordant, is the Is (Table 1.7).

Naturalism overlooks the qualitative step from the observation to the evaluation and makes a fallacious move from the former to the latter by which it reduces the latter to the former. Animism views the observation fully immersed in, and indistinguishable from, evaluation. Nihilism cannnot connect them.

The final way of thinking and acting listed here, the informationist praxiology, combines the affordance with valuableness in that informational systems can be cognisant of the valuable. The former is the precondition for the latter to emerge.

[d] John Collier and Michael Stingl speak even of the "good". They refer to living systems only. Here, the domain of the valuable is extended to all self-organising systems.

Table 1.7. The consideration of Is and Ought in informationist praxiology. The cognisantly affordant and the valuable.

		is	ought
con-flation	**reduction: naturalistic fallacy**	**the cognisantly affordant:** sufficient condition for the valuable	**the valuable:** resultant of the cognisantly affordant
	projection: animist fallacy	**the cognisantly affordant:** resultant of the valuable	**the valuable:** sufficient condition for the cognisantly affordant
disconnection: nihilistic fallacy		**cognisantly affordant**	**valuable**
		disparate takes	
combination: informationist praxiology		**the cognisantly affordant:** necessary conditions for a valuable	**the valuable:** an emergent from the cognisantly affordant

Wherever emergence takes place, a process/structure of the Third is the case. The valuable exists on a meta-level. It is contingent and it undertakes to shape the lower level for the valuable.

1.4.2 *Emergent reflectivity*

Praxiological valuableness is value-laden information for systems regarding which line of activity they are likely to follow and whether their response to some challenge is appropriate. Ontologically, information contains more than that. There is a difference between responsiveness as such (the capability of responding) and reflectivity (the capability of reflecting). Reflectivity includes evaluations but is more general and also includes cases in which values are not dealt with. How do the responsive and the reflective relate to each other (Table 1.8)?

Reflection is not a mechanical result of the response, as the reductionist position would hold. Reflection is also not the absolute cause of the response, as the projectionist position would argue. And reflection is not independent from the response, as the disjunctionist position would favour.

Reflections are emergents from responsiveness. The latter is, as potential, as non-being, the necessary condition for the former.

Table 1.8. The consideration of non-being and being in informationist ontology. The responsive and the reflective.

		non-being	**being**
con-flation	**reduction: mechanistic fallacy**	**the responsive:** sufficient condition for the reflective	**the reflective:** resultant of the responsive
	projection: teleological fallacy	**the responsive:** resultant of the reflective	**the reflective:** sufficient condition for the reflective
disconnection: existentialist fallacy		**the responsive**	**the reflective**
		independent existents	
combination: informationist ontology		**the responsive:** necessary condition for a reflective	**the reflective:** an emergent from the responsive

This is the position of UTI that is based on EST systemicity and POE integrity. Two perspectives are present: self-organisation that manufactures oneness, and a process of information generation that specifies self-organisation. This leads to a new definition of information. It refers to but modifies a triadic semiotics in the wake of Charles Sanders Peirce [2000]. It emphasises the intrinsic connection of self-organisation with negentropy after Edgar Morin [1992, 350, 368] and uses the term perturbation introduced by Humberto Maturana and Francesco Varela [1980]:

> [...] *relation such that (1) the order O built up spontaneously (*signans*; the sign) (2) reflects some perturbation P (*signandum/signatum*; (to-be-)signified) (3) in the negentropic perspective of an Evolutionary System* s_e *(*signator*; the signmaker).* [Hofkirchner 2013a, 171]

The order O is the relational order any self-organisation process adopts. The perturbation P is a perturbation among the participants of the self-organisation process; if those participants are autonomous systems, the perturbation is due to an influence on their activities from their environment; if those participants are elements of a common system, the perturbation is due to an influence from outside or inside their system. The negentropic perspective is the perspective to produce order, because entropy is tantamount to disorder; the self-organisation process reproduces an already built-up order or it produces changes towards an even higher order.

Systems are responsive and they are able to respond to perturbations by establishing a collective order or by changing their own order such that the response reflects the perturbations. Reflections are contingent based on the necessary possibility spaces of responses for both autonomous and suprasystems. Reflection is always reflection of, if not on, something. Accordingly, reflection is a meta-level entity: it is an emergent and thus a Third. Responsive systems undergoing a self-organisation process towards the establishment of a suprasystem are Firsts and together, as a Second, they can explore the possibility space open to them to counter some perturbation. They do this by building up the suprasystem order as the Third that reflects this very perturbation. Any single responsive system has a potential for a variety of responses to a variety of perturbations that can be responded to by changing the build-up of its order. Perturbations render the interaction network of the elements that are active in maintaining or changing the order of the system a Second. And such perturbations make every element a First.

When a system actualises a specific response through a self-organised act, it achieves a reflection of the specific perturbation in its own structure, in its own organisational relations by which it relates to the perturbation. This reflection is information for the sake of self-organisation. It is about the perturbation of self-organisation and construed by means of self-organisation.

Note here that information generativity is attributed to emergent systems according to the evolutionary class to which they belong. Human systems are not the only ones endowed with reflectivity (this would import the acceptance of Umberto Eco's threshold of semiosis applicable to the realm of human culture exclusively). Reflectivity is not only a feature of biotic, living systems (this is the threshold of Klaus Fuchs-Kittowski [1997] and the biosemiotics). Importantly, it also characterises physical, material systems in so far as they have the capacity to self-organise and are qualified to generate information. Emergent systems of any kind produce emergent information, albeit according to their kind.

Reflections come in three different manifestations, namely as cognition, communication or co-operation according to the Triple-C Model

[Hofkirchner 2013a, 184-196]. Cognition is information generated by every single informational system. Communication is information generated by the interaction of at least two informational systems. And, finally, information generated by integrating a sufficient number of informational systems with a suprasystem as information for and of this suprasystem is called co-operation.

Cognitive, communicative and co-operative reflections are each a Third on their own. But they harbour more. In cognition generated as a Third by one system, the Firsts and the Second of this Third are *intra*systemic. In communication, however, there are as many Thirds generated as there are systems taking part in generating communication (reflecting the communication of another system as a perturbation of its own Firsts and Seconds). Communication is *inter*systemic. Communication is generated by coupling cognitive systems. Co-operation is generated by entangling communicative systems. Co-operation is therefore *supra*systemic. It concatenates cognition and communication. In this concatenation, cognitive reflection of the participants in the suprasystem is a First, their communicative reflection a Second, and co-operative reflection a Third.

1.4.3 *Emergent semiosicality*

Ontological reflectivity, the information-generating capacity, is a relational property of self-organising systems. This is because information relates the system to perturbations stemming from its environment – other systems, co-systems, the suprasystem – or from its interior. Information is an emergent relation and therefore calls for epistemological consideration. What is the role of systemism with regard to the study of information, with regard to semiotics (the study of sign processes) (Table 1.9)?

One approach to the study of information departs from a third-person perspective. That approach considers information as an object that can be handled, as something material and thus measurable, and clearly as something external to human observers. That is how the mainstream (natural) science thread sees it. All semiotic considerations are reduced to what is believed to have been observed. The cultural semiotic approach starts from subjective human action that is considered immaterial and interprets information processes from a first-person point of view. Hereby, the subjective ideal is projected onto any appearances of information.

Table 1.9. The consideration of the apparent and the essential in informationist epistemology. The natural and the semiosic.

		apparent	essential
con-flation	**reduction: empiricist fallacy**	**the objective material:** sufficient condition for the semiosic	**the semiosic:** resultant of the objective material
	projection: hypostatic fallacy	**the natural:** resultant of the subjective ideal	**the subjective ideal:** sufficient condition for the scientific
disconnection: agnostic fallacy		**natural**	**semiosic**
		incommensurable knowledge	
combination: informationist epistemology		**the self-organisational:** necessary condition for a semiosically informational	**the semiosically informational:** an emergent from the self-organisational

The natural and the semiosic can be approached separately without reduction or projection if the third- and the first-person perspectives are disconnected. But this is to no avail.

The UTI approach makes the case for informationist epistemology when it combines the objective and the subjective into a subject-object dialectic, the material and the ideational into an emergentist materialist position. and the third- and first-person views into a perspective shifting methodology from the outside to the inside and from the inside to the outside. Self-organisation is the necessary condition for any semiosis, but semiosis goes beyond that. Information emerges from self-organisation and is a Third that differs from mere self-organisation.

This flexible framework approaches every information generation according to the evolutionary type to which the system pertains [Hofkirchner 2013a, 173-184]. Simple material systems show pattern formation, which is a stage of protosemiosic information generation in which all three semiosic aspects – the pragmatic, the semantic and the syntactic – are folded into one. The pattern is a protosemiosic sign that emerges based on a single self-organisation cycle. Living systems manifest code-making, in which a hierarchy of two self-organisation cycles produces a quasi-semanto-pragmatic sign on top of a quasi-syntactic sign. Human, social systems exhibit the constitution of sense. Here, the evolution of semiosis has unfolded to produce the full variety of pragmatic, semantic and syntactic signs.

1.4.4 *The Principle of the Co-Extension of Information and Self-Organisation*

Valuableness, reflectivity and semiosics are the main features of information emerging based on systemics.

Valuableness is the informational extension of the systemist functionality regarding synergy. Things have a value according to their contribution to synergy effects for both human and nonhuman systems. Such a value is both objective and subjective. It is information for the systems, but in order to obtain this information about the valuableness, a system needs to realise that value. This is an act of emergence. Values cannot be derived from the condition of the things or from the system's inclinations.

Reflectivity is the informational extension of systemicity. Systems act and react vis-à-vis their environment. The better they reflect their objective environment, the better they can realise their subjective goals. Reflections are an act of emergence. These do not follow logically from the systems' environment or from their goal-directed activities. The quality of the reflection depends on the system's cognitive, communicative and co-operative capabilities.

Semiosicality is the informational extension of systemist relationality. A sign is a relation that mediates between the system and its environment. The establishment of that relation is an act of emergence. It cannot be concluded from the state of the system in its environment. Any system is free to choose a sign of its own predilection.

A value, a reflection, a sign – all are Thirds. Their existence is not identical with, but grounded in, features of systemics – on emergent systems having functions, having activities to maintain or change their organisation and having organisational relations.

Thus, informationism can be defined:

Informationism. *Informationism is that* weltanschauung, *that conception of the world and that way of creating knowledge of the world that applies the Principle of the Co-Extension of Information with Self-Organisation, based upon the systemist Principle of Self-Organisation and the emergentist Principle of Unity-Through-Diversity.*

The **Principle of the Co-Extension of Information with Self-Organisation** states: there is, subjectively (as a matter of human systems' take) and objectively (as a matter of fact to be assumed by human systems regardless of whether or not they actually observe it)

(1) valuableness of systems of any kind for systems of any kind;

(2) reflectivity of systems of any kind;

(3) semiosicality in systems of any kind;

such that the valuable, the reflective, and the semiosic – thereby specifying the functional, the systemic, and the relational – are nested meta-level emergents, i.e., Thirds. Those Thirds represent information manifestations, actualised in ideations or materialisations by communicative networks as plural Seconds, in which cognitive systems participate as Firsts.

Any self-organisation generates information.

Part II

Steps to a Theory of the Social and of Social Information

Chapter 2

From Systemism and Informationism to Criticism

Social theorists have a tendency to write about the past as if everything that happened could have been predicted beforehand. [...]

Who knows? Perhaps if our species does endure, and we one day look backwards from this as yet unknowable future, aspects of the remote past that now seem like anomalies – say, bureaucracies that work on a community scale; cities governed by neighbourhood councils; systems of government where women hold a preponderance of formal positions; or forms of land management based on care-taking rather than ownership and extraction – will seem like the really significant breakthroughs, and great stone pyramids or statues more like historical curiosities. What if we were to take that approach now and look at, say, Minoan Crete or Hopewell not as random bumps on a road that leads inexorably to states and empires, but as alternative possibilities: roads not taken?

[...]

We can see more clearly now what is going on when, for example, a study that is rigorous in every other respect begins from the unexamined assumption that there was some 'original' form of human society; that its nature was fundamentally good or evil; that a time before inequality and political awareness existed; that something happened to change all this; that 'civilization' and 'complexity' always come at the price of human freedoms; that participatory democracy is natural in small groups but cannot possibly scale up to anything like a city or a nation state.

We know, now, that we are in the presence of myths.

– David Graeber, David Wengrow: The Dawn of Everything, A New History of Humanity, 2021 –

Part I demonstrated that the philosophical, the system theoretical and the information theoretical considerations of the new scientific paradigm – laid out in greater detail in the *Emergent Information* book – all underlie the same logic, the Logic of the Third. The next steps are to apply the principles of emergentism, systemism and informationism, in the same vein, to theories of the social and of social information, before, in Part III, social information technologies are theorised. This includes developing proper principles of this application in order to complete the framework of the new paradigm for the techno-eco-social transformations necessary to master the global challenges. Each step is a solidification of the previous steps on the path from the abstract to the ever-more concrete.

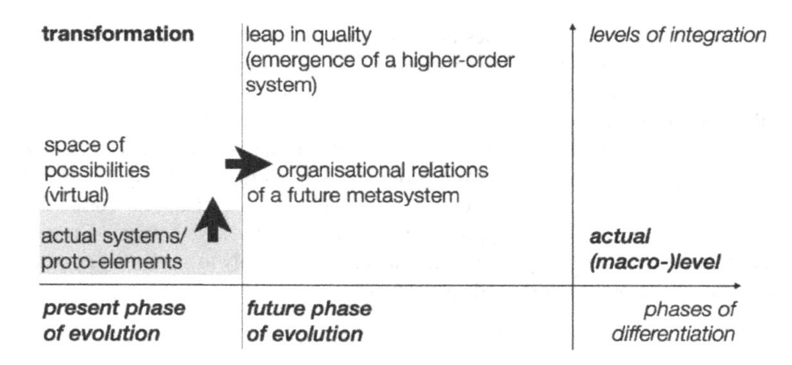

Figure 2.1.a. Transformation in emergent systems. The case of a qualitative leap – preparatory stage.

The question to ask is what is the concrete conclusion for the science of transformation arrived at in Part I. After all, that overview serves as a starting point for Part II. The take-home message is that transformation is the formation of a Third, recognisable in any emergent system. Transformation refers to any change of state of systems that accompanies any reproduction of the systems realised by their agents. This change might even outbalance the reproduction, yielding a qualitative change of the relations of organisation. A particular case is the emergence of meta-/suprasystems. A meta-/suprasystem is a step in the course of the evolution of systems that integrates differentiated systems on a higher level for the sake of synergy. This case is crucial for social evolution today (Figures 2.1.a and 2.1.b).

transformation	leap in quality (emergence of a higher-order system) space of new possibilities (virtual)	*levels of integration*
	actual organisational relations of the suprasystem	***actual macro-level***
past systems/ past proto-elements	actual elements of the suprasystem	***actual micro-level***
past phase of evolution	***present phase of evolution***	*phases of differentiation*

Figure 2.1.b. Transformation in emergent systems. The case of a qualitative leap – implementation stage.

Such a transformation is the showcase for the Logic of the Third. Actual systems of the same kind come, in their possibility space, upon a virtual option. Realising that option provides them all with organisational relations as a backbone of a future metasystem. All of them turn out to be Firsts as soon as they co-act with each other in a Second such that a Third emanates. They represent virtual proto-elements for the manifestation of a metasystem.

Another level appears. What was originally the macro-level of systems became, through their interaction, an intermediary level at the preparatory stage. This then shifted at the implementation stage to the micro-level of the suprasystem containing the former proto-elements as its new elements. The newly established macro-level of the suprasystem contains the new organisational relations. Those relations perfuse the whole new system piecemeal. The old systems are enabled und constrained to serve the new system they all share. Their task is to yield synergy for all of them by reproducing the suprasystem.

Part 2 of this book fleshes this transformation scheme out by rethinking the social and social information. Chapter 2 theorises social evolution and the evolution of social information from a critical perspective on general terms. Our understanding of the evolution of humanity is given a critical twist. The argument is that anthropogenesis – the evolution of humans – needs to be understood, first and foremost, as sociogenesis – the evolution of social systems. This is connected with noogenesis – the evolution of social informational systems – thereby positioning humanism as the philosophical, scientific and political idea of the human in a critical context. Sociogenetics and noogenetics are dealt with in two subchapters. They are followed by another chapter that tackles the same subject in more specific terms, focussing on the existential threats facing humankind since the mid-20th century.

2.1 Rethinking the Social: Critical emergentist Sociogenetics

The first subchapter develops cornerstones of a Critical Social Systems Theory (CSST). The point of departure here is the emergentist systemism of Evolutionary Systems Theory (EST) with its Principle of Self-Organisation.[a] This stands in contradistinction to what is widely known under the label of social systems, in particular originating in German-speaking countries connected to the work of Niklas Luhmann. This new approach was pursued not only by Bertalanffy or Bunge but also by members of Ervin Laszlo's General Evolution Research Group, among them Robert Artigiani [1991] and by representatives of Critical Realism, in particular Margaret S. Archer [1995; 2003; 2007; 2010; 2012] – who refers to works of US sociologist Walter F. Buckley [1967] – and her project group on Social Morphogenesis at the Centre for Social Ontology, including the economist Tony Lawson [2013] and the relational sociologist Pierpaolo Donati [Donati 2011; Donati and Archer 2015]. Of course, many other sociologists deserve mention; even if they do not explicitly share a systems approach, they have nevertheless contributed important insights to such a framework [Giddens 1984; Alexander 1995; Mouzelis 1995; Reckwitz 1997].

---

[a] That Luhmann's point of departure was Bertalanffy's General System Theory is just as poorly substantiated as Luhmann's claim to have adopted autopoiesis from Humberto Matura and Francisco Varela [Hofkirchner 2006; Fuchs and Hofkirchner 2009].

Importantly, there is an easy-to-make connection between such a social systems approach and Critical Theory tenets in the sense of the Frankfurt School and other Marxist-inspired or capitalism-criticising approaches.[b] Thus, this chapter bridges the gap between systemism and criticism. It does so by scrutinising the phenomenon of commons.

In social systems, social agents produce and use different material goods or different kinds of what is generally deemed an ideational good. Material and ideational goods exist in the form of relational goods [Donati 2014]. Any such good has the following properties:

> As an emergent entity, it is commonly produced and commonly used. It is produced through a common action although the actors can contribute to the common action in different ways. And although the use of the good by the actors might differ, it is, in principle, provided for common use. The good is produced by a co-operation of actors for the usage of actors who long for it, who demand it and who need it. So one can argue that every good is a common good that emerges from the relations of humans. For the same reason, it can be classified as a commons. [Hofkirchner 2014a, 69]

This way of revisiting commons when examining social systems can be termed "commonism" (in contrast to "communism"). The first mention of the term is traced back to US-American singer-songwriter Woodie Guthrie with reference to the Bible [Briley 2006, 35; quoted after Linebaugh 2014, 139]. Despite the differences in meaning, there is a connection to Marx's idea of communism. Commonism can be considered a generalisation of communism –

> [...] a generalisation that extends from material products via presumed natural givens, which far from being natural need more and more reproductive work by the economies and health care systems, to information including fields such as

[b] Though Bertalanffy might be associated with conservative views on social issues, the humanist value-ladenness of his system ideas, his realist stance and the methodology of his way of thinking allow a progressive interpretation of his oeuvre [Hofkirchner 2019]. This line is followed here.

> education or the arts. This summarises all phenomena mentioned under the term of the "commons". [Hofkirchner 2017b, 278] [c]

This connection becomes even clearer if the focus is on societal justice.

> The actors have a share in the added value when producing it and they share the added value when using it; but the share the actors have does not account for the added value produced nor does the added value produced account for how much the actors share. This problem of the lack of reciprocal accountability between costs by, and benefits for, individual actors is an argument against measurements of transactions and exchanges between individual or aggregate actors as the basis of measures to balance their rights and duties in a justifiable way; individual input to, and individual output from, the commons is rather a matter of collective action. And for that reason, the only principle of a humane organisation of production and usage of the commons that can be supported is, in general, "from each according to their ability, to each according to their need". [Hofkirchner 2014b, 80]

Since CSST relies on EST's emergentist systemism, thereby including criticism in an emergentist way, then commonism, as feature of CSST, is emergentist, too.

The detailed investigation of emergentist commonism follows the scheme of discussion introduced in Part 1. Commonism denotes, in that context, a social perspective (praxiology), a conception of the social world (ontology) and a social-scientific way of thinking (epistemology). A key aspect is communality emerging by the commons. According to these three aspects, the social is redesigned, remodelled and reframed. Redesigning yields common goodness, remodelling yields sociality as agency/structure dialectics, and reframing yields social relationality called structurality (Figure 2.2).

Since this subchapter introduces a concrete step from systems of any kind to social systems, the assignments to the different logics of argumentation – reduction, projection, disjunction – need to be replaced by those typifying main social science streams. The replacements hold for all sections of this and the next sub-chapter as well as the next chapter. Instead of naturalism, mechanicism, and empiricism, individualism is used

---

[c] In fact, a lecture held by Slavoj Zizek in the first decade of our millennium in Vienna first made me aware of that idea. I merely completed the generalisation by applying systems thinking.

here to characterise the logic of reduction in order to connect with well-known methodological individualism; instead of animism, teleology, and hypostasis, sociologism – introduced by Bunge [2003, 164] – is used here to signify the projective logic; and instead of nihilism, existentialism, and agnostics, anarchism is used to designate the disjunctive logic with reference to the anarchist theory of knowledge after Paul Feyerabend.

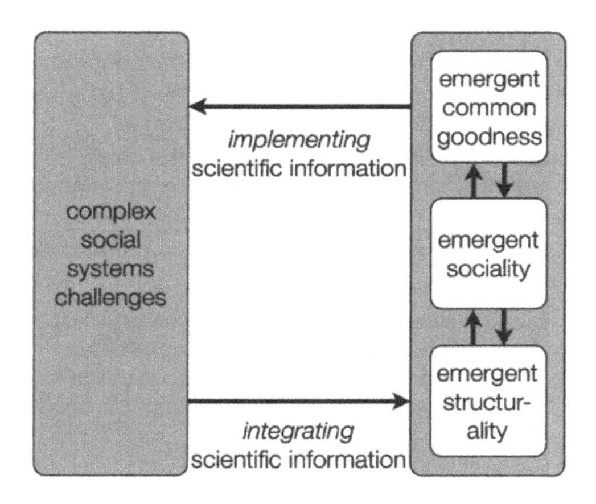

Figure 2.2. Critical emergentist-systemist sociogenetics. The emergent commonist content of the new paradigm.

2.1.1 *Redesigning the social: the emergence of common goodness*

The social "Is and Ought" draws on the systemist processual and objective functional. The first possible stance is reductionism, which in the vein of Bunge can be called social individualism [2003, 113]. It prioritises the individual over and against society or any other social system that he or she is an element of. This potentially leads to the prevalence of private concerns and self-interests. A long tradition alleges that, as a rule, something good results from the self-concerned active for the good of the common sociality. This tradition can be classified as mere ideology[d] (Table 2.1).

Table 2.1. The consideration of Is and Ought in sociogenetic praxiology. Actors' actions and the common good.

<table>
<tr><th colspan="2"></th><th>is</th><th>ought</th></tr>
<tr><td rowspan="2">con-
flation</td><td>reduction:
individualistic
fallacy</td><td>the private:
sufficient feasible condition for the common good</td><td>the common good:
desirable resultant of the private</td></tr>
<tr><td>projection:
sociologistic
fallacy</td><td>the private:
feasible resultant of the common good</td><td>the common good:
desirable condition for the private</td></tr>
<tr><td colspan="2" rowspan="2">disconnection:
anarchist fallacy</td><td>private</td><td>common good</td></tr>
<tr><td colspan="2">disparate takes</td></tr>
<tr><td colspan="2">combination:
sociogenetic
praxiology</td><td>the private on loan:
necessary feasible condition for a good of the commons</td><td>the good of the commons:
a desirable emergent from the private on loan</td></tr>
</table>

The reversal of social individualism – sociologism – does not reduce the social to the individual, but projects the former onto the latter. It refers to the prevalence of social relations, which in sociology configure the structure of the social system over and against the individual or agency. The latter two are constituted by social agents called actors in sociology. The good is projected accordingly. Projections yield problematic positions such as Amitai Etzioni's controversial Communitarianism because they hypostatise the role of duties [1995].

The simple negation of both projectionism and reductionism is to deny that one relatum is subsumed under the other relatum, to which here an anarchist fallacy is attributed[e]. It cuts individuals free from society (or any

[d] The best-known example is from economics: market imperfections falsify such individualistic assumptions.

[e] The notion of anarchism, used here as a label for indifference in the way of thinking, allows another association to social movements. In a restricted sense, the notion also appears to be adequate in that it conforms with libertarian varieties of anarchism. Importantly, however, it fails to do so in the case of altruistic anarchism.

social system), or cuts agency free from structure, and promotes independent views of the processes of the individuals and the structured good.

2.1.1.1 *Cycles for the good or for the bad*

Sociogenetic praxiology integrates the two relata without impeding their differentiation. It connects individual agency and the common good in that it construes individual agency as commoning[f] activity, the common good as a good of commons, and their relation as emergence of the latter from the former together with a dominance of the latter over the former. The common good that is key to a good society in the sense of Aristotelian Eudaimonia is defined as the good of the commons. The private that is something that deprives society of a good of the commons must be understood as being on loan from society only.

Goodness anticipates, ideal-typically, a self-propelling cycle: by commonly producing the commons, actors produce relational commons for common usage (Fig. 2.3.a):

> Happier individuals conduct a better life together such that they bring about the formation of relations that improve the promotion of the formation of happier individuals. [Hofkirchner 2017b, 279]

This perspective includes the systemism of EST and criticism.

The systemist perspective is due to the emergent functionality of synergy:

> The rationale of every system is synergy. Because agents when producing a system produce synergetic effects, that is, effects they could not produce when in isolation, systems have a strong incentive to proliferate (Corning 2003). In social systems synergism takes on the form of some social good. Actors contribute together to the good and are common beneficiaries of that good – the good is a common good, it is a commons. [Hofkirchner 2014c, 121]

---

[f] That term can be found in Linebaugh [2014, passim] and on https://blog.commons.at/commons/was-ist-commoning, meaning, approximately, taking care of something together.

> So, since any (self-organised) system is a system by virtue of the synergy it supplies through its organisational relations, any social system is a social system by virtue of the common good it supplies through its organisational relations. The common good is the social manifestation of synergy. Social actors play a major part in adding to those values and themselves share the added values in turn. They are producing and using the commons in common, they are commons' co-produsers. [Hofkirchner 2017b, 280][g]

The cycle in the figure is to be read clockwise, starting from the bottom: commons co-produsers, when enjoying the co-action of commoning, can share in the production of relational commons and can share in their usage. The cycle is a virtuous circle in which commonness spirals up.

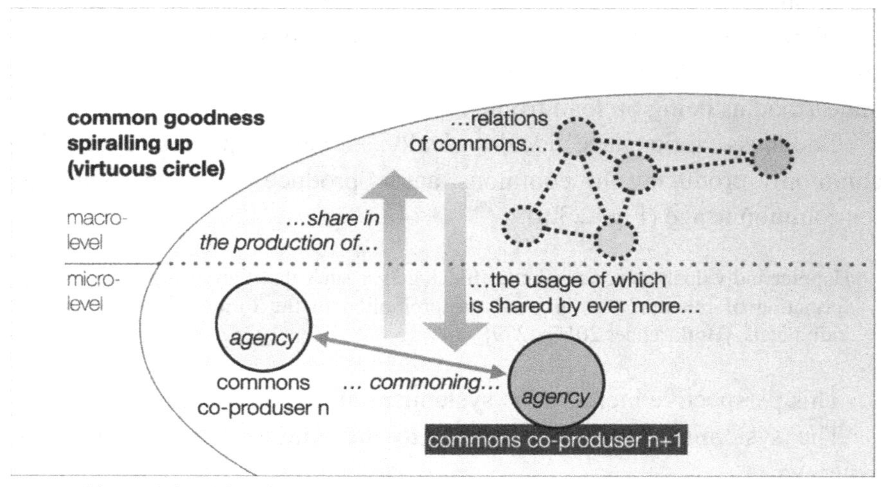

Figure 2.3.a. The virtuous circle of social systems. Disclosure of commons.

The criticist perspective specifies what signifies a good society. It signifies a society in which human emancipation is enshrined according to Karl Marx's "*categoric imperative to overthrow all relations* in which man is a debased, enslaved, abandoned, despicable essence" [1844]. It is the objective of a critique of critical social sciences and their theories to

---

[g] The terms "produser" (and "produsage") arose with modern ICTs that turned the users of social network sites into the produsers of content of the latter [Bruns 2006]. They are taken here as a metaphor for any social systems production and usage on a generic level.

provide scientific knowledge for emancipation. "Science is partisan", wrote Kurt P. Tudyka [1973, 25 – my translation] in the aftermath of the 1968 protest movements in his introduction to political science. "Criticism is a method oriented towards the recognition and sublation of contradictions" [1973, 9 – my translation], a statement that can be generalised across the range of all social sciences. Before the Positivism debate that took place between the Frankfurt School Critical Theory and Critical Rationalism in Germany, the ideology of value-free science excluded considering the application context in research and viewed values as a breach of scientific rules. In the vein of Tudyka, instead of mirroring reality, which suggests the immutability of what would need to be changed, the task of social science should orient itself towards conceptualising a concrete utopia that transcends the empirically undesirable by referring to what is really possible [24-25]. Science becomes critical if it measures current reality against the possible reality that is desired.[h]

Thus, the critical social-systemist praxiology specifies the conditions for human emancipation. If humans are debased, enslaved, abandoned, despicable beings, its root cause is the improper cultivation of the commons, the exclusion of produsers from contributing to and benefitting from the commons (Fig. 2.3.b).

---

[h] Since the founding of the General System Theory by Bertalanffy and others, it goes without saying that systemism is itself normative too. Its inherent normativity forms a new, scientifically grounded *weltanschauung*. As communication expert Mark Davidson pointed out, Bertalanffy's approach is inherently value-laden because it emphasizes becoming aware of the mutual dependencies of systems, elements, subsystem and suprasystem, and the hierarchical relations of the architecture so as to include the collective consequences of the individual actions taken [Davidson 1983; 2005, 30]. "Bertalanffy looked for an Is that suits its supervenience by Oughts that are grounded on the Is". "Despite the widely agreed assumption that Ought cannot be derived from Is without additional premises that contain values, a closer look at that issue might do better justice to the intricate relationship between them. Is can be seen as a necessary but not sufficient condition for Ought, and Ought can, in turn, be seen as a contingent emergent on a given Is" [Hofkirchner 2019, 268].

If commons are enclosed, actors are excluded from a fair share in both producing and using those commons such that they are forced to compete with each other. That is the vicious circle of commonness spiralling down. The greater the enclosure of the commons, the more competition for inclusion, and the greater the competition, the more the enclosure.

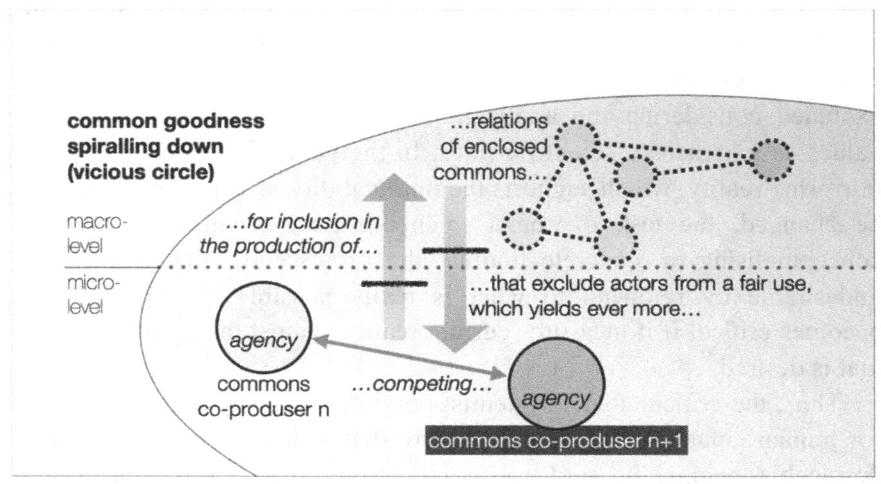

Figure 2.3.b. The vicious circle of social systems. Enclosure of commons.

Both social systemism and criticism reflect the Logic of the Third on their own. Combined, by establishing the praxiological meaning of the good, emancipatory society of the commons, they bring to the fore the social system in the true sense of its adjective. This commonness plays the role of the Third. Individual actors are the Firsts, who organise themselves into a network as the Second and can, eventually, bring about the Third.

2.1.2 *Remodelling the social: the emergence of sociality*

Common goodness does not only stand for the good that emerges with social systems. It also points to the real "mechanisms" – as social ontologists call them – that underpin the emergence of goodness and form the dynamics that build up and maintain social systems. This dynamics is grounded in the dynamics of systems in general, as EST shows. Again, different attempts have been made to model that, some of which are doomed to failure (Table 2.2).

The individualistic fallacy views the actor's imagined options in the space of possibilities – her or his virtuality – as a sufficient condition for social transformations – the social actual. Ontological individualism means, as prime minister Margaret Thatcher once put it, "there is no such thing as society. There are individual men and women, and there are families" [Thatcher 1987]. The sociologistic fallacy makes the social (trans-)formation determine the option imagined by the actor. Ontological sociologism is holistic (in the negative sense of the word), "holism underrates agency and overrates bonds" [Bunge 2003, p.112]. The anarchist fallacy cannot decide between what is conditional and what is transformative, what is virtual and what is actual. Ontological anarchism disjoins individuals from society or agency from structure – they viewed as being independent from each other – and promotes a dualistic division.

Table 2.2. The consideration of non-being and being in sociogenetic ontology. Agency and structure.

		non-being	**being**
con-flation	**reduction: individualistic fallacy**	**the individual:** sufficient agential condition for the social	**the social:** organised resultant of the individual
	projection: sociologistic fallacy	**the individual:** agential resultant of the social	**the social:** sufficient organised condition for the individual
disconnection: anarchist fallacy		**individual**	**social**
		independent existents	
combination: sociogenetic ontology		**the co-/intra-active individual:** necessary agential condition for a meta-/suprasystemic social	**the meta-/suprasystemic social:** an organised emergent from the co-/intra-active individual

2.1.2.1 *Agency/structure dialectics*

Sociogenetic ontology applies the notion of the possibility space to the social in that it brings agency and structure together as follows. The social space provides possibilities of (trans-)formed structures that might be realised. In accordance with the systemic self-organisation loops, a social (trans-)formation comes into being in two scenarios: if and when the agency of actors induces structural changes – either a new structure or changes in the structure – and if and when those actors have previously been under some environmental conditions or under structural conditions of their social system that evoke pursuant action.

This ontology is also in line with emergentist systemism and social criticism. It is not about "substance", erroneously stating that the social is a "substance" that differs in kind from "substances" of pre-human reality (biological or physical "substances").

According to EST in the wake of Bertalanffy, the object of social sciences is not a substance at all. The social is distinct from the pre-social through a different and emergent organisation. It is always the organisational relations that account for the quality of a system. Here, Bertalanffy – who reconciled the debate between physicalism and biologism (as we would term the debate between mechanicism and vitalism today) – anticipated Bruno Latour's critique of a substantialist view in sociology. However, Bertalanffy's answer to such a view was a system theoretical one that does not reflect the direction of Latour's actants and networks but rather aims in the direction of the so-called new materialism or agential realism [Latour 2005; Barad 2007 and 2012].

> The object of social sciences is not a particular substance but rather social systems, which means the social relations that organize matter in a different way than in realms of prehuman, biotic, and physical matter. [...] In short, the object of social-scientific inquiry comprises (a) the actors that co-act to such an extent that the social system is reproduced or transformed, (b) the social relations that emerge from and, through the provision of constraints and enablers, dominate their interaction, and (c) the interplay between actors and social relations, in which the actors remain in the space determined by social relations or transgress it and in which the social relations turn up as intended or turn out to be unintended consequences. This is in line with a general definition of systems that includes (a) elements, (b) organizing relations, and (c) their interplay, which is self-organization. [Hofkirchner 2019, 268]

Social self-organisation is shown in Figure 2.4. The actors as bearers of agency in the social system are depicted here such that they resemble the architecture of the social system – internally they possess two levels that are connected through a bottom-up and top-down cycle. The actors *co*-act when producing social organising relations for a nascent meta-system and *intra*-act when reproducing or transforming the social relations of the mature suprasystem they belong to. In any case, they act such that those relations provide them with ever better means for a good life (Fig. 2.4).

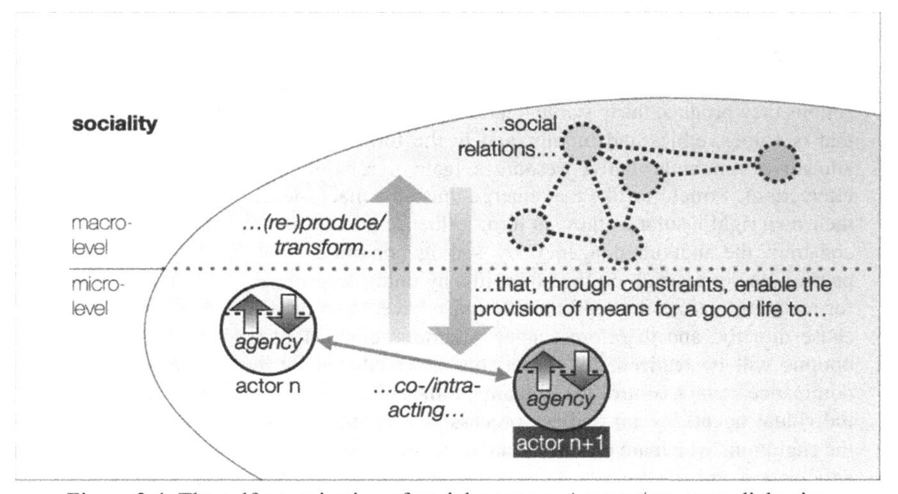

Figure 2.4. The self-organisation of social systems. Agency/structure dialectics.

This also agrees with what, according to Tudyka, is a second precondition of social criticism put by Marx [1852] into the famous words:

> Men make their own history, but they do not make it as they please; they do not make it under self-selected circumstances, but under circumstances existing already, given and transmitted from the past.

Social critique considers that the link between agency and structure is a contradiction, i.e., a dialectic relationship. Such a relationship is said to exist once both relata are opposed to each other, depend on each other, and cannot be replaced with the other without simultaneously replacing the mode of the relationship. In this dialectic relationship of agency and structure, the latter is a Third because it relates the actors to each other. It orders the behaviour of actors by opportunities that enlarge and/or diminish their options.

In terms of EST-based CSST, agents and structure are modelled as being coupled in a feed forward and a feedback loop. This self-organisation either reproduces or transforms the system, including possible metasystem transitions. More than twenty years ago, the idea of a dialectic of social self-organisation cycles took on the following formulation [Hofkirchner 1998, 29-30]:

> There are two levels. At the micro-level the elements of the system, namely agents, are located. They carry out actions, and by the interplay of the fluctuating individual actions they produce fairly stable relations among them which, in the form of rules, that is values, ethics and morals, and in the form of regularities which concern allocative and authoritative resources, gain a relative independence from the interactions. Structures like that emerge thus on a macro-level, where they exist in their own right insofar as they, in turn, influence the agents. On the one hand, they constrain the individual agency by setting conditions that limit the scope of possibilities to act and, on the other, just by doing so provide it with the potential for realizing options it would not otherwise have. In so far as the structures do not cause directly, and therefore cannot determine completely whether or not these options will be realized, for the actions are mediated by the individual agents, dominance cannot control the outcome, either. The structures are inscribed in the individual agents by an endless process of socialization and enculturation, but the engramms which are produced in the individuals serve as cognitive tools for the anticipation and construction of ever new actions which may or may not obey the rules and accept the values and recognize the ethics and follow the morals, and which may or may not fit the regularities and renew the allocative and authoritative resources and thus may or may not reproduce the structures. Either way, interaction reflects upon the conditions of its own emergence and may consciously be directed at the structures in order to maintain or alter them. In this sense only, that is, because in their recursive actions the agents refer to the structures, these structures play the dominant role in this relation of bottom-up and top-down causation. Nevertheless [,] none of the relations in this causal cycle leads to plain results. Each influence has consequences which due to the inherent indeterminacy cannot be foreseen. By this, and only by this, qualitative change is possible.

2.1.2.2 *The architecture of society as nested systems*

Social systems that are known as societies contain internally differentiated subsystems and are based on originally different non-social, natural systems which they integrate. Societies form a multiplex of levels such that individual actors participate in systems at different levels at the same time. Social systems at those levels can be ordered along a specification hierarchy as follows in a top-down manner [Hofkirchner 2014c, 121-122][i]:

(1) **Cultural systems**. Since culture is widely deemed tantamount with what distinguishes social evolution from natural evolution, it is no surprise that what is termed cultural systems here populates the highest level; that realm is characterised by actors that, in their cultural roles, produce rules to define their self-fulfilment; what can be called cultural commons are the societal relations that define what (a) good is in a self-fulfilled good life.

(2) **Political systems.** Political systems are positioned below this; actors in their political role produce and execute decisions to regulate their self-determination (political decisions form part of cultural rules); the political commons are the societal relations that condition the decision process on the conduct of a self-determined good life.

(3) **Economic systems.** Economic systems are on the next-lower level; actors in their economic role produce and allocate resources for their self-reliance (economic resource dispositions form part of political decisions); the economic commons are the societal relations that condition the distribution of means for a self-reliant good life.

(4) **Eco-social systems.** Apart from these social subsystems, eco-social systems originate from integrating natural systems into society; actors in their eco-social role produce adaptations to, or of, the natural environment to support their self-preservation (eco-social adaptations form part of the economic resource dispositions); the eco-social commons are the societal relations that condition the interplay of the material objects and the material subjects in a self-preserving good life.

(5) **Techno-social systems.** And techno-(eco-)social systems at the bottom level form the basis of that hierarchy; actors in their techno-eco-social role produce and use scientific-technological innovations to enhance and augment their self-actuation (techno-(eco-)social achievements form part of the eco-social adaptations); the techno-(eco-)social commons are the societal relations that condition the

---

[i] Ideas of emergent nested systems were applied to urban systems, albeit in a slightly different hierarchy, by Christian Walloth [2016].

material ways and means of human activities, physical tools and procedures in a self-actuating good life.

In this sense, society is a social system that encapsulates different social systems, including systems of non-social, natural origins. Each system on the levels of the cultural, the political, the economic, the eco-social and the techno-eco-social is a Third in its own right when viewed in relation to the actions (the Second) of the respective actors (the Firsts). Moreover, any system on a higher level is also a Third because any step upwards is a process of change in quality. It crosses the boundary of the specific quality that systems share on a certain level towards an integration with a higher quality that does not exist on the lower level. In contrast, any step downwards is towards a differentiation into qualities that are continuously shaped by the higher quality.

> The order results from the fact that every sublevel is a concretisation of the preceding upper level: culture is the level where rules are defined; politics is the level where regulations (specific rules) are decided; economy is the level where resource regimes (specific regulations) are deposed; ecology is the level where the natural infrastructure, the eco-structure (specific resource regimes), is designed; and technology is the level where the technical infrastructure, the techno-structure (specific natural infrastructure), is designed. [Hofkirchner 2017b, 282]

Importantly, the general feature of such nested systems is that the lower levels "exhibit a faster pace of interaction between the actors and a faster pace of system transformations as compared with the higher levels. […] the techno-structure has the highest rate of innovation and culture the lowest one" [Hofkirchner 2017b, 282].

2.1.2.3 *Socialisation/sociation, individualisation/individuation*

Sociality, emerging by social self-organisation, is a Third that exhibits a balanced confluence of two counter-rotating processes (see again Fig. 2.4): the upward process of "soci(alis)ation" and the downward process of "individu(alis)ation". Those processes are an instance of the dialectics of integration and differentiation.

> Agents that differ from each other bring about the formation of a structure that integrates; this structure catches up with the differentiation at hand and, at the same

> time, conditions a new differentiation of the agents which brings about a new integration: ever more differentiated agents bring about the formation of an ever more integrated structure that conditions the formation of ever more differentiated agents. There is divergence and convergence in one. That is the line along which evolution can make progress. [Hofkirchner 2014c, 128]

Soci(alis)ation describes the increase in the "socialness", "commonness" or "communitarity" as the integrity of social systems. Individu(alis)ation, in turn, describes the increase in individuality as an inviolable property of actors. Through the first process (on the left-hand side in Fig. 2.4), actors interact in an associative, i.e., co-active or intra-active way by which they integrate with the social system, with society. According to this process, they produce those social relations (on the right-hand side) that enable them to become different individuals in an individual way, to develop capabilities to actualise themselves. The more the actors can individuate/individualise, the more they can "sociate"/socialise and drive a virtuous circle. In contradistinction, the less they can find their individual way due to constraints imposed by the social relations they dislike and dismiss, the less they can "sociate"/socialise and the more they are forced to dissociate through counter-action. This leads to ever more restrictive, opposing social relations – a vicious circle. Both circles clearly involve the disclosure or enclosure of the commons (Fig. 2.3.a and b). The disclosure orients towards the ideal-type of a balanced sociality in harmony with individual development. The enclosure induces a tension between both processes. In such cases, socialisation outbalances the possibilities of individuation, leading to authoritarian collectivism, or individualisation outbalances sociation, leading to anti-social individualism [Hofkirchner 2014c, 128-129].

2.1.3 *Reframing the social: the emergence of structurality*

The emergence of sociality as a human feature is not possible without the emergence of structurality in the sociological sense. Structure (social organisation implemented by social relations) is not only ontologically important, making it as real as actors or their acts, but is also epistemologically significant.

What is the difference between understanding the structure of a social system and understanding the behaviour of the social system's actors?

Again, the epistemological enigma of sociogenesis can be illuminated by contrasting the different framing approaches. The individualistic approach is a reductionist fallacy. The empirically observable behaviour of individuals, groups, classes etc. is viewed as a sufficient condition that structures society and theoretically explains the behaviour. The sociologistic approach, however, reduces the structure of society to behaviour, making the structural devoid of theory. This approach is therefore a projectionist fallacy, projecting the structural back to the behavioural such that the latter lacks empirical facts. The anarchist approach, while not conflating the behavioural and the structural, fails to connect what it looks upon as being incommensurable.

Table 2.3. The consideration of the apparent and the essential in sociogenetic epistemology. Behaviour and social relations.

		apparent	essential
con-flation	reduction: individualistic fallacy	**the behavioural:** sufficient empirical condition for the structural	**the structural:** empirical resultant of the behavioural
	projection: sociologistic fallacy	**the behavioural:** theoretical resultant of the structural	**the structural:** sufficient theoretical condition for the behavioural
disconnection: anarchist fallacy		**behavioural**	**structural**
		incommensurable knowledge	
combination: sociogenetic epistemology		**the superficial behavioural:** necessary condition for a theoretical structural	**the yet-to-be explored structural:** an emergent from the empirical behavioural

2.1.3.1 *The observable and the unobservable*

Sociogenetic epistemology can compensate these shortcomings by accepting the emergentist, critical combination of both the behavioural and the structural: the behaviour is empirical, and though it is superficial, it provides the necessary conditions to explore a theory of the structure. Such a theory is an emergent.

Pierpaolo Donati and Margaret S. Archer invoke "a general theory of social relations [...] from the point of view of Relational Sociology" [Donati and Archer 2015, 18-19]. They sketch a "relational order of reality" [19]. For them, to state that certain entities are in relation

> [...] is an ontological expression that has three analytical meanings: (i) it says that, between two (or more) entities there is a *certain distance* which, at the same time, distinguishes *and* connects them; (ii) it says that any such relation *exists*, that is, it is real in itself, irreducible to its progenitors, and possesses its own properties and causal powers; and (iii) it says that such a reality has its own *modus essendi* (the modality of *the* beings who are *inside* the relation which refers to the internal structure of the social relation and its dynamics) and is responsible for its emergent properties, that is, relational goods and evils. [Donati and Archer 2015, 18]

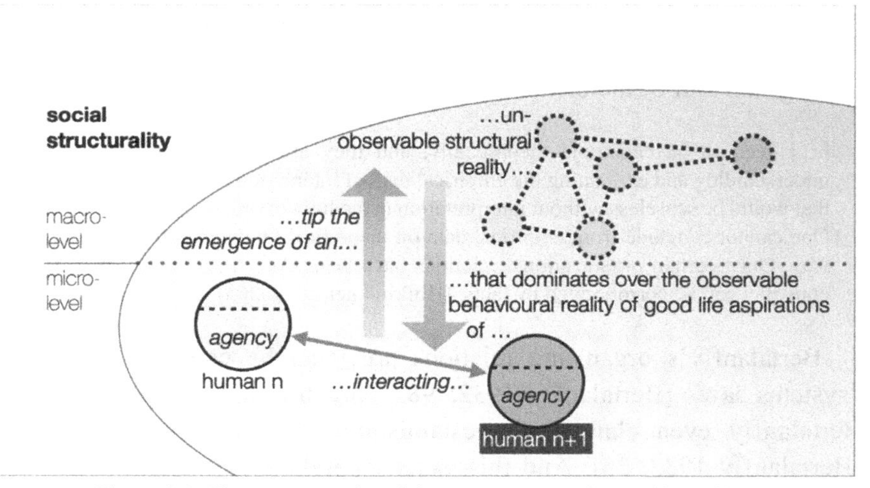

Figure 2.5. The structure of social systems. An unobservable reality.

The structurality of social systems is illustrated in Figure 2.5. The empirical behavioural part is located on the micro-level where agents interact and can be observed. When they co-act (or intra-act) to aspire a good life, they tip the emergence of the social-relational part on the macro-level (represented by the upward arrow from the micro- to the macro-level). This creates a structural reality that exerts dominance over the observable part of reality in that it relates the interacting agents (downward arrow from the macro-to the micro-level) but is itself unobservable and can be explored only by theoretical conjectures.

Bertalanffy's achievement in laying the grounds for systemism was his awareness of such an epistemological task. He understood that his turn to relations of organisation – although initially focussed on organismic systems of the biotic realm – is a theoretical issue. "Science is not a mere accumulation and catalogue of facts. It is a conceptual order we bring into facts" [Bertalanffy 1953, 238]. He emphasised the difference between description and explanation, between perception and a "system" of concepts. Only the latter, Bertalanffy is convinced, can put science in a position to deal with "laws" making theory indispensable for science [Bertalanffy 1932, 22-25]. The relations of organisation can be tipped by a co- or an intra-action of the elements of a system and manifest themselves in the behaviour of each element through constraints and enablements. Nonetheless, they themselves are not observable. In contrast to the behaviour of the elements, they

> [...] need to be construed theoretically, and they are necessary in promoting understanding and explaining the empirical data of interaction and behaviour – data that would be senseless without interpretation in the light of organizational relations. One cannot conclude from empirical data on theoretical knowledge in a deductive way. Once certain organizational relations have been hypothesized; however, the construct can be corroborated by facts. [Hofkirchner 2019, 265]

Bertalanffy's organising relations are what he called in 1932 the "systems law" [Bertalanffy 1932, 98 – my translation]. At that time, Bertalanffy even claimed the establishment of "theoretical biology" [Bertalanffy 1932, 25]. And this exactly parallels the situation we find today in sociological and overall social science research investigating social systems. Accordingly, the criticism both with regard to the "biological" and the "social" is very similar.

> Every kind of systems thinking is close to the idea of always incorporating a Third in that it is a feature of such thought to look upon every object as immersed in an overall systems context. Evolutionary Systems Theory treats any event or entity as process, that is, as the result of a process that propels evolution of systemic interconnectedness. That way it works on a meta-level that provides a Third to weave the red thread among the components at the empirical level. This symbiosis

of criticism and systemism becomes obvious when focusing on the question of the behaviour ("Verhalten" in German) of social actors and the social relations ("Verhältnisse" in German) between them. This question has been attracting attention anew with Pierpaolo Donati's paradigm of Relational Sociology (2011). Behaviour can be investigated empirically. Not so social relations. The latter can be identified only through theoretical endeavours. Social relations are, so to speak, that which is essential for behaviour, that is, what is common because it is necessary, and may be labelled "lawful" in this sense. Social relations appear in concrete behaviour. In systems terms, they are the enablers and constraints of the actions and interaction of the actors. They determine, in a way, the behaviour of the actors. In this way, behaviour can be understood by referring to its underlying social relations. But no behaviour can be explained by resorting to the actors or agency alone. Enablements and constraints are relational, they are structural in nature, not agential. They are the Third that relates actors, and individual agency realises only possibilities that are undergirded by social relations. [Hofkirchner 2015, 99]

Marx wrote that "all science would be superfluous if the outward appearance and the essence of things directly coincided" [Marx 1894]. All critical social-scientific investigation must be aware of, and try to bridge, the incongruence of appearance and essence. Its task is to reconstruct the link between the two. Only by "theoretically reconstructing the historical totality and relinking the single aspects to that Third, social science is able to give meaning to empirical findings and to provide scientific understanding" [Hofkirchner 2015, 99]. Tudyka writes:

Criticism gains power when it can put the object in the context of societal totality, thus recapturing it from illusive empirical isolation and demonstrating its historical society-wide character. When considering single aspects in isolation, criticism becomes lost practically and cognitively, and surface manifestations squelch critical thought. [Tudyka 1973, 12 – my translation]

2.1.3.2 *Typology of social relations*

The societal totality is not the context of phenomena on the same level as that totality. It is the structure of the social whole. Gaining knowledge about the structure of society is important for an analysis that can inform practice. Critical theory orients towards the sublation of social antagonisms. Antagonisms are a special type of social relations that can theoretically govern any social system (Table 2.4).

(1) **Antagonism.** Antagonisms demand unity without diversity. They demand uniformity in an absolutist claim. "Antagonisms arise from constellations of positions, appearing in the subjective behaviour of social actors and materialising in objective social relations, such that a gain for either side is a loss for the other. Hence the term 'zero-sum games'" [Hofkirchner 2015, 105]. Those positions are oppositions that are mutually exclusive. They are contradictory. It seems that a conflict can be settled only by the success of one side, which makes both sides prone to violence. There are two incorrect framings of antagonisms.
   (a) **Universalism.** One is universalism, which is tantamount with egalitarianism in the sense of levelling down any difference for the sake of identity. It is a negation of any variety.
   (b) **Particularism.** The other is particularism, which is known as fundamentalism. It is an imposition of one particular difference as a common identity on all the other varieties.

(2) **Agonism.** Another type of social relations is so-called agonism [Mouffe 2013]. Agonisms demand diversity without unity in a relativistic sense of plurality, propagated by postmodern ideology. "Agonisms are constellations of positions, appearing in the behaviour of social actors and materialising in social relations, such that no gain for either side has any implications for the other" [Hofkirchner 2015, 106]. Those positions can co-exist with each other; they are compossible in the sense of Gottfried Wilhelm Leibniz, they are not contradictory, they are contrary, a juxtaposition of every variety in its own right, indifferent as to their differences. According to Mouffe, antagonisms should be transformed into agonisms, a contest without violence.

(3) **Synergism.** Critical theory and systems thinking give rise to a third type of social relations that can claim hegemony for the betterment of society. It is the Third that sublates antagonism as well as agonism: social synergism is in a position to reconcile the contradictions and fructify the contraries or indifferents for the sake of the social unity through diversity. It is

> [...] a constellation of identity and difference in social affairs that is deemed to be the ideal one [...]: as many differences as possible [...] produce as much identity as may be necessary [...]; the differences are identical in as much as they identify themselves with the identity they commonly produce, while identity, in turn, is differentiated as long as differentiation does not lead to the disintegration of identity. In game theory that is termed "non-zero-sum games" in "win-win situations". [Hofkirchner 2015, 105]

This critical relationalism postulates a convergence of mutually supportive (pro-)positions that complement each other in a composition of varieties that make themselves fit for the whole. On the one hand, social unity (the Third) is required such that not every interaction of social varieties (the Second) might be appropriate to promote social unity. On the other hand, it is in the sovereignty of any social variety (the Firsts) to abide by what is required for the united sovereignty.

Table 2.4. Types of social relations. How to reconcile identity and difference.

		social relation	**identity and difference**	
ab-solut-ism	**universalism (egalitarianism)**	**antagonism** demanding uniformity	**contradictoriness:** conflict of mutually exclusive (op-)positions	**negation** of any variety
	particularism (fundament-alism)			**imposition** of one variety on any other variety
relativism (postmodernism)		**agonism** demanding plurality	**compossibility:** co-existence of (in-)different positions	**juxtaposition** of every variety in its own right
critical relationalism		**synergism** demanding unity through diversity	**complementarity:** convergence of mutuallly supportive (pro-)positions	**composition** constituted by varieties made fit

Given this typology of social relations, and given that there is a seamless flow between antagonisms and agonisms in both directions, one task of critical relationalism is to provide answers to the question of how to turn antagonisms into lasting agonisms. This is supported by Mouffe and is accepted here as an intermediary step. A second task is to turn agonisms into lasting synergisms, which is the Third [Hofkirchner 2015, 106-111].

This typology is associated with the virtuous and vicious circles of sociality. Antagonisms and agonisms are bound to the vicious circle of sociality. They fracture it into individualism, according to which actors on the one hand fight each for their own part of the commons, and into collectivism, on the other, where all actors are apathetic receivers of non-transparent commons transfers. The synergism is key to the virtuous circle in which commoning soci(alis)ation drives commoning individu(alis)ation and vice versa.

2.1.4 *The Principle of Commonism*

Critical emergentist sociogenetics – the theoretical approach to the investigation of sociogenesis, the origin and evolution of social systems – is a cornerstone of a Critical Social Systems Theory (CSST). Such sociogenetics is based on the new paradigm of systemism and emergentism. It is the first step in applying its principles to the human world. It applies the principle of emergent systems and concludes that anthropogenesis – the origin and evolution of humans – so strongly reflects the emergence of social systems that humans and social systems cannot be divorced. This has implications for humanism, the notion of what is human and what is humane.

Humanism revisited, update I. Humans are dependent on living in a society, and societies are dependent on being maintained by humans. This intricate relationship is an absolute condition of the human(e). And this relationship is a relation of commons, as explained above in sections 2.1.1 to 2.1.3. The following considerations all revolve around the commons and justify considering the commons as a core feature of humanity: the common goodness as the emergent good of the commons; sociality as the emergent dialectics of agency and structure, whereby agency refers to the development of commons, and structure refers to the conditions of agency; and structurality as the emergent reality of unobservable relations of commons. The commons are the most precious thing humans hold objectively. Accepting the following ideas is therefore a necessary ingredient of humanism: the commonist idea is humanistic, and, even more so, commonism provided by the new paradigm is the update of humanism in its entirety.

It can be defined:

Sociogenetics. *Sociogenetics is that critical social elaboration of* weltanschauung, *that critical conception of the social world and that critical social-scientific way of creating knowledge that applies the criticist Principle of Commonism, based upon the systemist Principle of Self-Organisation and the emergentist Principle of Unity-Through-Diversity.*

The **Principle of Commonism** states: there are, regardless of whether or not thematised, associated with the advent of social systems,

(1) the emergence of the common good in the form of the commons that represent social synergy;

(2) the emergence of sociality through a dialectics of agency and structure according to which actors co- and intra-act; in doing so they build up, reproduce and transform the social relations of the social meta- or suprasystem; accordingly, in turn, the social relations constrain the actions, by which – and as long as – they enable the production and provision of commons and thus preserve the social system;

(3) the emergence of structurality as an objective social-relational condition of the commons production and provision in which the subjective conduct of the actors is embedded; that conduct cannot be observed like the empirical behaviour of the actors but must be explored theoretically;

such that common goodness, the social, and the structural to-be-explored – thereby specifying the functional, the systemic, and the relational of the systemist approach – are instantiations in a hierarchy of levels. At the same, each is an emergent meta-level on its own, that is, a Third that represents an essential property of social self-organised systems, actualised in ideations or materialisations by the interaction networks of the actors as plural Seconds. In those Seconds, the participating actors – humans or other social systems – represent single Firsts. Any social system organises itself via the social organisation carried out by their proto-actors or actors and as long as they do so.

Humanism is instantiated as commonist.

2.2 Rethinking Social Information: Critical emergentist Noogenetics

The first step here was to elaborate the Principle of Commonism as the paradigmatic content of sociogenetics in order to take the leap from systemism to social criticism. Based on that, the second subchapter applies informationism to social criticism and provides building blocks for a Critical Information Society Theory (CIST). The approach involves reconstructing the subjective side of social evolution as the noogenesis of *Homo creator* in a creative universe. Noogenesis is the origin and evolution of what the Soviet Russian father of biogeochemistry, Vladimir Ivanovich Vernadsky, called in the 1930s the noosphere [Vernadskij 1997]. This was prompted by exchanges with Edouard Le Roy, who was a scholar of Henri Bergson ("L'évolution créatrice"), and with Teilhard de Chardin. His concept of the noosphere, however, differed in that it was compatible with, and based on, (natural) scientific evidence. Vernadsky's concept ties in with the concept of the biosphere introduced by the Austrian geologist Eduard Suess, who is known for planning and constructing the potable water pipeline for Vienna in the 19th century. In line with Suess, Vernadsky defined the biosphere not as the mass of living beings but as a sphere that is produced by living beings through their interventions into the originally inert geosphere that drives ecological cycles of chemical substances. Analogous to the relation of biosphere to geosphere, Vernadsky theorised the evolution of a noosphere as a result of anthropogenic impact on the biosphere and on the geosphere. Humans started from the outset through work, science and technology (i.e., by means of social information) to create a sphere of reason and thought, a sphere of the noetic. Hence the term "noosphere". By noogenesis, the human race and its information generation capabilities has effected impacts of geological dimensions on a time scale that is orders of magnitude faster than the biotic impacts. This can clearly be interpreted from an emergentist informationist-systemist point of view (Fig. 2.6).

In Fig. 2.6, noogenesis is one of the latest terrestrial transformations. Only three major steps are referred to here.

(1) **Geosphere.** The first step is the formation of planet Earth; it is the phase of planetary accretion through self-organisation of physical

systems that have rudimentary capacities for information generation, establishing the level of a geosphere.

(2) **Biosphere.** The second step encompasses the spread on Earth of living systems with a wider range of information capacities than mere physical systems; by reorganising physical systems, they establish the level of a biosphere (the biosphere extends into the geosphere and the relations of physical systems are overruled by biotic relations, whereby original physical systems become bio-physical elements of the biotic systems).

(3) **Noosphere.** The third and final step, noogenesis, is the start and spread of even more differentiated information-generating social systems that reorganise biotic systems and establish the level of a noosphere (the noosphere extends into the biosphere, whereby it overrules the biosystems relations so as to replace them by social structures and turn biosystems into socio-biotic subsystems; by doing so, the latter make the bio-physical elements of living systems into an *agens* of socio-bio-physical sub-subsystems).

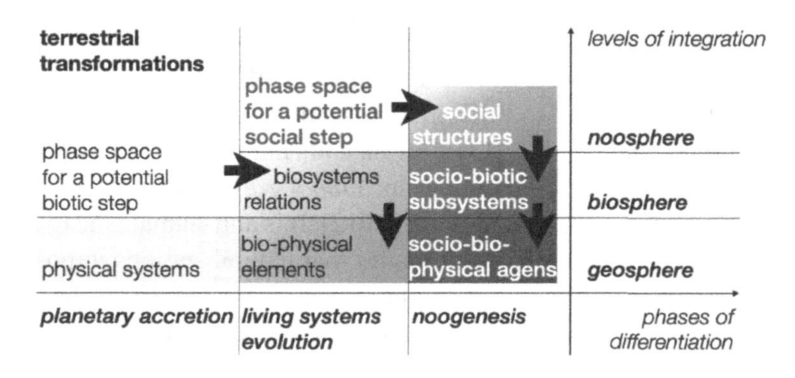

Figure 2.6. Noogenesis as terrestrial transformation.

Each step represents a qualitative leap in transformation, originating by emergence from self-organisation of agents at lower levels and representing a specific Third that shapes the lower levels. The evolution of self-organising and information-generating systems has led on Earth from physical over biotic to social systems with physical, biotic and social, noetic, information [Hofkirchner 2013a]. Noogenesis is part of that evolution leading from nature to society. Noogenetics designates a humanistic, scientific and evidence-based theoretical investigation of the genesis of noetic processes.

Noogenetics is inherently socially critical because it does not reduce all planetary interactions to cybernetically self-sustaining natural circular flows. Nor does it qualify humankind as terminal cancer for the planet, as was done by the Gaia hypothesis [Lovelock 1979 or 2019]. And it does not reduce the question of human evolution to the search for geological artefacts of human impacts that would stand the test of geological time categories as was intended by the neologism "Anthropocene" [Crutzen and Stoermer 2000]. Those who coined that term meant it in a positive, but hubristic, sense and did not flinch from propagating geo-engineering as if such measures were merely a normal continuation of the development of creative human mind. Today, a majority in favour of the term are inclined to mean it in a self-destructive, pejorative sense. Both senses lack a true humanistic stance: they cannot do justice to the social character of anthropogenesis and to the noetic features its social character implies.[j]

Noogenetics can build upon the theoretical findings of the first subchapter to embrace social criticism. The current subchapter explicates that sociogenesis has been endowing humans with those noetic features that make them prepared to fulfil their possible destiny of a co-operative species [Bowles and Gintis 2011]. Nowak and Highfield [2011] even refer to super co-operators, representing an *animal rationale* [Schnädelbach 1992] with a human social mind [Pagel 2012]. It is anti-humanistic to deny such a destiny based on alleged obstacles that natural settings impose on humans or with regard to alleged future trans- or posthuman technology options. Moreover, such a denial is a theoretical misjudgement of the new quality of the social (compared with the natural) and its informational evolution and disregards empirical facts.

Commonism is the crucial starting point. The following features are indications of the objective existence of an ambivalence of the

---

[j] Vernadsky, who died in January 1945 was, in 1941, convinced that Hitlerism would be defeated because it violated the basic regulations of the planetary noosphere. Despite the breach of socialist principles in the Soviet Union, with its millions of slaves, its state would still be able to enforce the moral encirclement of the wartime enemy [Krüger 1981, 93-99].

(anti-)social: the two possible circles concerning common goodness (and badness) either disclosing or enclosing the commons; the two possible circles concerning sociality (and anti-sociality) of either a harmonic socialisation through sharing the commons or of individualism; and the two possible circles concerning structurality (and anomy) either towards unity bolstered by the sharing of the commons or towards the retention of agonisms and, even worse, towards the aggravation of antagonisms. There is a possible "good" and there is a possible "evil", and humans are beings that can take either path, as French sociologist and philosopher, Edgar Morin, formulated with his concept of *Homo sapiens and Homo demens* [Morin 1979],

> [...] having both original goodness and original vice tightly intertwined [...]. This ambivalence must be taken in stride. It involves weaknesses, miseries, deficiencies, mercilessness, goodness, nobleness, possibilities of destruction and creation, consciousness and unconsciousness. [Morin 1999, 114]

Humans, however, are equipped with subjective capabilities to inform themselves about the choice they can make. Noogenetics ascribes those noetic means to humans.

Humans can generate social information and make use of it. They are creative. The term *Homo creator* goes back to Nicholas of Cusa, a German cardinal, a philosopher as well as mathematician and scientist, a proponent of the Renaissance humanism.[k] Bertalanffy studied him when developing his ground-breaking ideas on systems. He took Nicholas's idea "*ex omnibus partibus relucet totum*" ("each part reflects the whole" – an idea anteceding Leibniz's monadology) as a point of departure. According to Nicholas, the totality is an infinite universe as "*coincidentia oppositorum*". By that it also inheres the good and the evil. In Nicholas's late pantheistic mysticism, God is no longer the coincidence of all opposites, but rather

[k] Later, that term was used in the 1950s and 1960s by Jewish Swiss philosopher, Michael Landmann, and German sociologist and ethnologist, Wilhelm Mühlmann, who had been a member of the NSDAP. Those conceptions are clearly based on different *weltanschauung* settings. I myself used the term in [Hofkirchner 2003] without any reference. In the explanations that follow, I refer only to Nicholas of Cusa's concept as interpreted by Bertalanffy and others because of its striking closeness to systemic thinking.

over and above all opposites – more than the totality of all things, specified by Nicholas as "*non aliud*" ("Nicht-Anderes", "non-other" – my translation) [Bertalanffy 1928, 15-28]. Interpreted in terms of the Logic of the Third, God is on another level, the level of a Third. Furthermore, God is characterised by creativity, and humans, as "*secundus deus*" ("second God"), can, by imitating God, participate in the plentifulness of creative nature with their own "*vis creativa*", their human creativity, which they use to produce conjectures about the infinite universe [Borsche 2017, 31-34]. Thus, again interpreted in the sense of the new paradigm for today, humans, as *Homo creator*, are able to create a multitude of Thirds when endeavouring to gain knowledge – that knowledge is fallible and needs to be repeatedly overhauled in the creative universe.

If Thirds exist objectively and can be conjectured subjectively by social information, then this entails two possibilities: the objective realisation of the "good" as a Third that prevails over the "evil", and the subjective self-making of *Homo sapiens* as a Third over and above *Homo demens* (which does not mean that *demens* is eradicated). The Logic of the Third calls for a switch from the dominance of *demens* to the dominance of *sapiens*, the containment of *demens* as a continuous task of noogenesis.

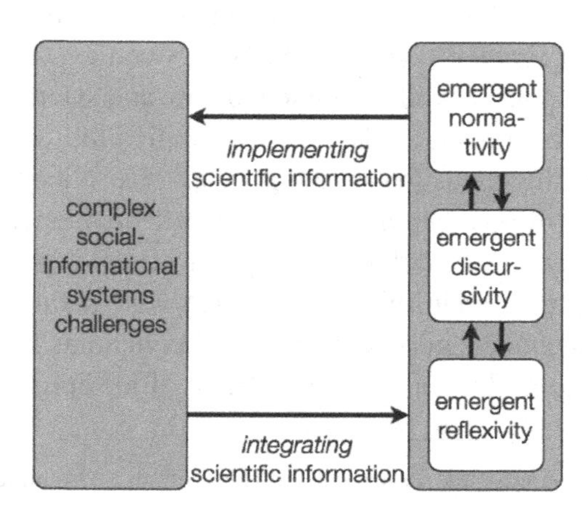

Figure 2.7. Critical emergentist-systemist-informationist noogenetics. The emergent *Homo creator* content of the new paradigm.

The emergentist concept of *Homo creator* is the critical content of social informationism in praxiological, ontological and epistemological respect, the content of critical noogenetics, based upon sociogenetics (CSST) and informationism (UTI). Given these relationships, it provides the basis of what can be understood as the so-called information society (CIST).

Recapitulating the Triple-C Model of UTI helps put the following three sections in context. Figure 2.8 shows the informational build-up of any self-organising system (from the physical to social).

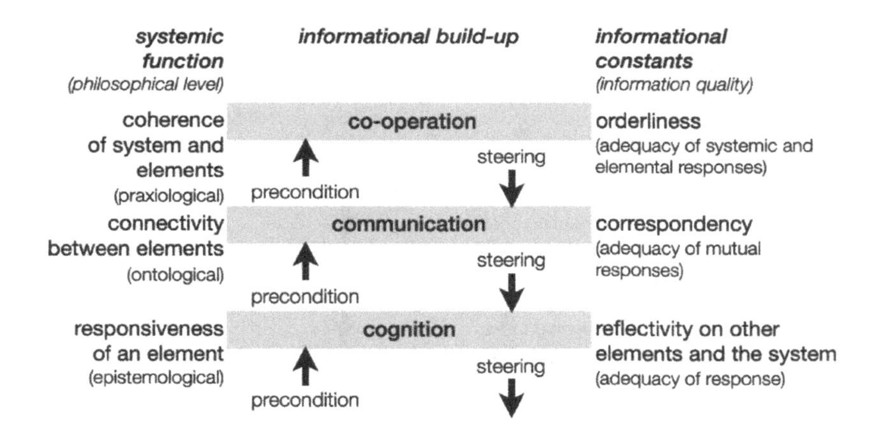

Figure 2.8. Informational build-up of self-organising systems.

Cognition, communication and co-operation are the fundamental information processes. Each is situated at a specific level that is a precondition for the level above and a steering of the level below, fulfilling a specific systemic function.

(1) **Co-operation.** On the topmost level, the function fulfilled is to maintain the coherence between the system and its elements, the elements and their system. It is enacted by the informational constant of orderliness – the adequacy of systemic and elemental responses to their mutual perturbations[1].

---

[1] Perturbations are any internal or external influences on the unit of self-organisation, including agential activities due to information generation.

(2) **Communication.** On the intermediate level, the function is connectivity between the elements; this is fulfilled through informational correspondency – the adequacy of their responses to their mutual perturbations.

(3) **Cognition.** On the bottom level, responsiveness of any element is a key property that is achieved through the reflexivity of any element such that it is able to reflect (on) other elements as well as on the system as a whole – the adequacy of response to perturbations.

Fig. 2.9 illustrates how these provisions translate into informational constants in social systems.

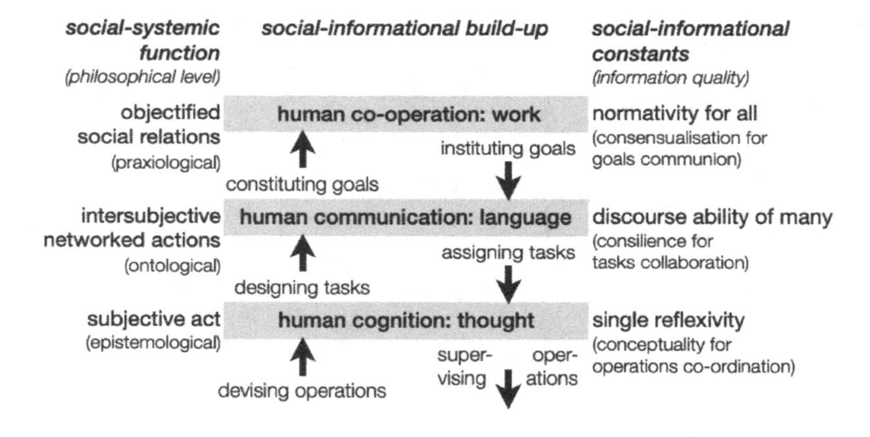

Figure 2.9. Social-informational build-up in social systems.

To describe it again from a top-down perspective:

(1) **Work.** As to coherence, human co-operation can be subsumed under the term "work". Work that emerged with humans is objectified in the structure by the validity of goal consensualisation[m] regarding goal setting and goal achievement. Here, in a system-theoretically, social-scientific sense, the term work denotes the process that is directed at

[m] A term used by Hermann Haken in a lecture I attended in Vienna in the 1990s. He said that he was criticised by social scientists for the naming of his "Slaving Principle" and that he meant rather a process of consensualisation of particles.

the build-up and development of the societal system. Normativity is valid for all actors that engage in work.[n]

(2) **Language.** As to connectivity, human communication is bound to emergent language. By discourses that can elaborate mutual understanding, intersubjective action networks can devise and distribute tasks – a process termed intersubjectification – in which many actors collaborate according to the goals. Consilience[o] is a *sine qua non* for effective collaboration.

(3) **Thought.** As to responsiveness, human cognition is distinguished by the emergence of thinking. Single actors use reflexivity to achieve judgements to co-ordinate, plan and control all operations. This enables them to determine their subjective act in relation to collaboration. Finding consent is actualised by using concepts, not percepts [Logan 2007]. This process by which a single actor attempts to resonate with supra-individual social perturbations is called subjectification.

Normativity, discursivity and reflexivity are the social information constants of work, language and thought. This is the topic of the following sections.

2.2.1 *Redesigning social information: the emergence of normativity*

The social-informational "Is and Ought" is based upon the social "Is and Ought's" emergent common goodness and on the informational "Is and Ought's" emergent valuableness. The noetic "Is and Ought" is built by converting the social interplay of privatising actions and the common good, as well as the informational interplay of cognising affordances and the evaluation of those affordances, into a new interplay of intentions and dedication. These are characteristic of human, that is, social information processes.

---

[n] Since every human is assumed to do so, normativity can even be understood as the outstanding characteristic of humans versus animals [Niedenzu 2012].

[o] That term became popularised after Edward O. Wilson used it in the title of his book on Sociobiology [1998].

How do old ways of thinking and the new paradigm perform in construing that interplay (Table 2.5)?

The conflationist variants fallaciously either reduce or project the dedicative to the intentional. They do this according to the ways of thinking of individualism or sociologism, respectively. The first cannot work because intentions of different actors do not automatically lead to a common dedication instituted in society. The second cannot work either because an institute of common dedication of actors need not prejudice the intention of any of the actors.

Table 2.5. The consideration of Is and Ought in noogenetic praxiology. Actors' intentions and consensual dedication.

		is	**ought**
con-flation	**reduction: individualistic fallacy**	**the intentional normative:** sufficient condition for the dedicative	**the dedicative:** resultant of the intentional normative
	projection: sociologistic fallacy	**the intentional normative:** resultant of the dedicative	**the dedicative:** sufficient condition for the intentional normative
disconnection: anarchist fallacy		**intentional normative**	**dedicative**
		disparate takes	
combination: noogenetic praxiology		**the shared normative intentional:** necessary condition for a consensual dedicative for a communion	**the consensual dedicative for a communion:** an emergent from the shared normative intentional

The disconnective variant cannot provide a good explanation of how intentionality and the societal institute of dedication could come together.

In lieu of the old paradigm, noogenetic praxiology combines the two sides in that the sharing of intentions of actors is the necessary condition for the emergence of a consensual dedication. A consensus in sharing intentions to reach a certain goal turns that goal into one to which all those who entered the consensus are dedicated. It is valid for all of them. It makes their group a communion. This yields normativity.

The new paradigm stresses the role of shared intentionality to a degree that it becomes key to the understanding of the origin of (wo)man.

2.2.1.1 *Two steps of noogenesis*

The Russian psychologist Alexei N. Leontyev's [1981, 210-2012] activity theory is known for its classical hunter and beater example. It illuminates how co-operation in the new collective action – the production and distribution of food through hunting – caused communicability as well as cognitive activity to become functional. Human actions are viewed as being distinct from animal behaviour in that they do not consist solely in the direct satisfaction of biotic needs but are mediated by a detour to the social context. Actions make sense because they are embedded in commonly designed chains of actions.

Michael Tomasello was co-director of the Max Planck Institute for Evolutionary Anthropology in Leipzig from 1998 to 2018. There, he conducted experimental research to compare the faculties of chimpanzees and bonobos with those of children, before he recently published a theory of human ontogeny [2019]. That work cleared the ground for an understanding of anthroposociogenesis and its noetic implications from an evolutionary perspective [1999; 2008; 2009; 2014; 2016]. By hypothesising from empirical findings, he distinguishes two decisive steps in the evolution of social life:

(1) **Joint intentionality.** First, he postulates a shift in co-operation from individual to joint intentionality due to enforced interdependence.
Individual intentionality is the point of departure. It is present in chimpanzees as well as in the common ancestors of chimpanzees and humans, perhaps as far back as in hominins about 6 million years ago. These taxa exhibit co-operation, for example in foraging situations, and favouring kin and other conspecifics over themselves. Nonetheless, the individuals seem to be driven by self-interest and are rather competitive. Once a group succeeds in gaining food, they eat without co-operative features. They might be referred to as monads without the social whole typical of modern humans. There is no need to take common goals into consideration – no need for thinking on a level beyond the actual ego-centric perspective.

Personal self-regulation is the rule.

Things change about 400,000 years ago in conjunction with altered environmental conditions. Early humans, living as hunters and as gatherers, developed joint intentionality. They extend "their sense of sympathy beyond kin and friends to collaborative partners" and build so-called dyads for co-operation. Such dyads are driven by second-person morals, that is, agreements for a common way to exploit food sources at least between two partners. They develop "a common-ground understanding of the ideal way that each role had to be played for joint success" and create "original socially shared normative standards" that recognise "that self and other were of equivalent status and importance in the collaborative enterprise" [Tomasello 2016, 4]. This satisfies the need for acknowledging a common goal, for understanding that the partner shares the goal and that both are committed to act according to its achievement in a fair manner. Personal self-regulation gives way to a social self-regulation.

This social innovation in which co-operation partners evaluate each other (by inserting social factors) triggers an accelerated biotic evolution in early humans.

(2) **Collective intentionality.** Tomasello conceptualises the second shift as a shift in co-operation from joint intentionality to collective intentionality due to another stimulant for even closer interdependence.

This shift appears in early humans in tribes about 150,000 years ago, ushering in modern humans, when groups grew bigger. The important step is the development of triads for co-operation, driven by "a kind of scaled-up version of early humans' second person morality" – "a kind of cultural and group-minded, 'objective' morality" [Tomasello 2016, 6]. With the advent of culture, the interdependence that necessitates co-operation reigns "not just at the level of the cooperating dyad, and not just in the domain of foraging, but at the level of the entire cultural group, and in all domains of life" [Tomasello 2016, 85]. Modern humans are "cooperatively rational": they factor in "that helping partners and compatriots [...] is the right thing to do"; "that others are equally as real and deserving as themselves"; and "that a 'we' created by a social commitment makes legitimate decisions for the self and valued others, which creates legitimate obligations with moral identities in moral communities" [Tomasello 2016, 160]. Social self-regulation becomes normative self-governance: since then, modern humans ask, "What ought I to think? And what ought I to do?" [Tomasello 2019, 21]. They need to know that any person belonging to the same group culture can be expected to share the same morality and, thus, responsibility for everyone else. From the Logic of the Third point of view, a meta-level is constructed in social reality – any group member can imagine the whole of the group, the roles taken, his/her own as well as others' replaceability. Such an idea was anticipated by US philosopher George Herbert Mead's "generalised other" [Mead 1934] – a Third extending the dyads to triads.

Due to that second social innovation, the biotic evolution of humans becomes dominated by social evolution. Noogenesis begins to develop in its own right.

For Tomasello, "human morality is a form of cooperation" [Tomasello 2016, 2]. His "interdependence hypothesis" states that individuals responded to necessities of ever-growing interdependencies between them. The first step was because of climate changes, the second step because of demographic changes. His "shared intentionality hypothesis" states that individuals developed the ability to share their intentional states to build a plural-agent "we". Altogether, this explains how morality came about. This led him to answer the question: "Which is the most plausible candidate for being the factor that, in non-human biotic systems, actualised a potential and fed back to the non-human biotic systems so as to transform them into a new – the human – system?" [Hofkirchner 2016, 285]. The new form of co-operation is what began to distinguish humans from the animal ancestors back then and from the great apes today. This comes as no surprise if co-operation is considered a Third. Co-operation is situated on the macro-level of the systems and it is the organisational relations where changes become manifest. Insomuch as nascent humans seized the opportunity to develop their new forms of cooperation, given exterior or interior challenges, they demonstrated their creative ability from the outset. The conclusion is: they created themselves and are still creating themselves.

2.2.1.2 *Typology of the normative*

Normativity here denotes "the property of social systems to regulate human behaviour through norms, values and interests, through morals and through ethics. Morality is at the core and ethics is a meta-level phenomenon" [Hofkirchner 2016, 279]. These noetic categories are defined as follows [Haug 2004; Hörz and Hörz 2013; Klaus and Buhr 1974; Sandkühler 1990; Weingartner 1971]:

> A norm is a collective expectation that represents an imperative to act in a determinate way and can be formulated as statement such as "you ought to do X (in circumstances Y)". An addressee of the imperative may or may not act accordingly.
>
> A value is a collective attribution that represents the meaningfulness of an object and can be formulated as statement such as "Z is true", "Z is beautiful" or "Z is good". A given subject may or may not share any of these three evaluations.
>
> An interest is a propensity for an individual intention that depends upon a collective entitlement, such that an actor is inclined to act in a way that is determined by the position the actor assumes and by the rights and obligations that go hand in hand with that position. An interest represents the directedness of the actor towards an object in accordance with the actor's location in the social relationships involved. It can be formulated as a statement of the kind, "I intend to appropriate Z in a way that is my proper right or obligation".
>
> Morality is that part of normativity that implies a reference to goodness. Morals are about goodness. A norm is moral or morally relevant, if and when the expected action is considered as good, in terms of degrees of the good or of the evil. A value is moral or morally relevant, if and when the property that is attributed to something is "good", "more or less good" or "evil". An interest is moral or morally relevant, if and when the intention is good, more or less good or evil, and the entitlement on which it is based is good, more or less good or evil.
>
> Ethics is the philosophical, scientific, or every-day reflexion of morals or moral issues. [Hofkirchner 2016, 279-280]

Ethics is thus in the position to influence morality, which is on an ideational object-level.

2.2.1.3 *Consensualisation qualities*

Figure 2.10 sketches human co-operative information build-up. The build-up has three levels. On the left-hand-side, Figure 2.10 depicts the instances of different information faculties (representing different kinds of action), on the right-hand-side, it shows the respective different information qualities (and the criteria by which they can be measured).

Every higher level is distinct from the level below through an emergent quality, which makes it more complex. Equally, every higher level is dependent on the lower level and exerts a dominance over that lower level in that the latter is shaped for the purpose of the former by conditioning its boundaries:

(1) **Monads.** Using Tomasello's steps of hominisation, the least complex information instances are the individual monads. Monads experience their world, including their social world. In fact, they are embedded in their social world and have the faculty to co-ordinate collective action if they prefer so.

(2) **Dyads.** Dyads are inter-individual, thus going beyond the monads in that they have the faculties to agree on mutual acts of collaboration. In fact, under the condition of triad existence, dyads are already intra-social, which reinforces those faculties.

(3) **Triads.** The social triads are the most complex information instances. They can share intentions in order to act as a body – a collective action that constitutes sense.

Co-operation is all about consensus, and the consensus appears in different information qualities. The fully-fledged quality of consensualisation emanates with normativity on the most complex level:

(1) **Normativity.** Norms, values, interests are adopted by triads to implement system objectives. That is the quality of consensualisation that creates normativity. Normativity can be measured: the default value is being moral; it can, however, also drift to being immoral.

(2) **Collegiality.** On the dyadic level, consensualisation takes the form of sharing findings of affordances of objects that would satisfy needs or wishes. Collegiality is its quality, which is set to fairness over and against unfairness.

(3) **Concernedness.** The preference of monads for acting in common depends on their social concernedness. If they are socially concerned, the information they create has an allocentric touch – that is, an orientation towards the others – and not an ideocentric one – towards his or her own self.

Overall, normativity requires collegiality on the next-lower level, and collegiality requires social concernedness to implement consensus on all levels.

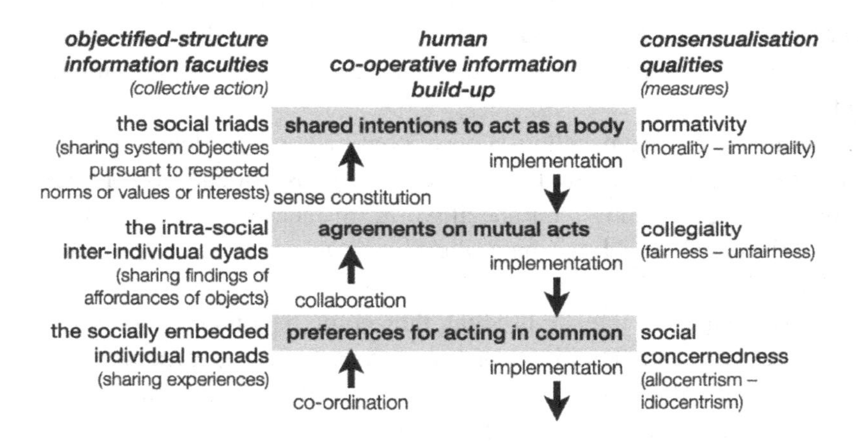

Figure 2.10. Objectified social-informational build-up of work in detail. Faculty instances and work settings for normativity.

2.2.1.4 *The morals of the commons*

The critical twist enters noogenetics when the subjective moral values are bound to the objective commons. The factual and the normative are related in real world.

> On the one hand, it is a fact that there are objective functions that need to be carried out in a social system and they are carried out by enactments of the actors' agency. On the other hand, it is a norm when actors are collectively expected to act in a determinate way; it is a value when actors collectively attribute a determinate meaning to an object; it is a moral norm when actors are collectively expected to act in a good way; it is a moral value when actors collectively attribute goodness to an object [...].

> Morality enters the stage because of the commons. That commons are generated and furnished by any social system is an objective fact. That commoning is doing good, is a moral norm for actors. That commons are a good thing, is a moral value for actors. For that reason, any enactment of social agency that contributes to a better fulfilling of the systemic function of yielding the commons is made a moral norm since any actor who is member of that system ought to do so for the sake of the good life, which is, at the same time, for the sake of the perpetuation and improved performance of the system, and any objective systemic property that facilitates the fulfilling of that systemic function is made a moral value since every member of the community of actors of that system will evaluate it as highly conducive to achieving good life, which is, at the same time, conducive to the functioning of the system. Thus morality is conveyed by the objective functions and properties of the social system. That is, what serves the objective functioning of the social system, appears to the actors as laden with morality. [Hofkirchner 2017b, 282-282]

As in any self-organising system, two different selves are involved in social systems as well: the system as a self in its own right, alongside the elements as selves, i.e., society and actors respectively, seen in a third-person perspective. The setting of the good-society structure is connected with the setting of good(-life) individuals; the objective function of the societal self is connected with the objective function of the agential selves. Those agential selves attribute in first-person perspective to the objective functions of society moral norms, values and interests that are here termed collective because the addressee is a collective. They attribute to the objective functions of individuals moral norms, values and interests that, in contrast, are termed individual here because the addressee is an individual.

In the present context, the conceptualisation of the moral items is meant in a universalist sense because it is based upon systems properties that apply in a universal way. Thus, the items are intended to represent universal human norms, values and interests. Universality does not forego concretisations due to further social-informational development, as will be shown in Chapter 3.

Morals in the overall system. Regarding the overall social system (society) and its actors (Table 2.6.a), the societal objective function of these "re-creative" systems [Hofkirchner 2013a, 113-114, see also Figure 4.12., 111] can be called self-production to denote social reproduction and social transformation to create the commons. This objective function appears to the individual actors as collective moral items of justice (fairness), cohesion, inclusion. For the actors, this means social compatibility, i.e., a just distribution of the commons in a society forming a cohesive whole by being inclusive of all its parts. For the sake of such a good-society setting, the actors are called upon to materialise those collective moral items.

The implementation of the societal objective function depends on the implementation of objective functions of individual actors. This can be called self-invention, denoting the transcendence of their selves while retaining their identity during any action for the sake of creating themselves by means of the commons and by contributions to the commons. They ascribe to the enactment of self-invention the individual moral item of dignity. To them, this means self-worth, a sense of self, but also of togetherness, underpinned by consciousness and conscience. This involves conditioning by the commons and, in turn, conditioning the commons.

Table 2.6.a. Objectified social-informational build-up in work. Societal and agential objective functions as well as collective and individual moral items in general.

	good-society structural setting		**good-life agency setting**	
	objective function of the societal self	**collective moral norms, values, interests**	**objective function of agential selves**	**individual moral norms, values, interests**
the overall social system	**self-production** (creating commons)	**justice (fairness), cohesion, inclusion** (social compatibility)	**self-invention** (creating themselves by enjoying, and contributing to, commons)	**dignity** (self-worth by enjoying, and contributing to, commons)

Morals in the subsystems. Concerning the social subsystems and their actors, the societal objective function of self-production is differentiated into three different objective functions (Table 2.6.b): into self-expression of the cultural subsystem, by which cultural commons are provided; into self-government of the political subsystem, by which political commons are provided; and into self-sustenance of the economic subsystem, by which economic commons are provided. The commensurate collective moral items of social compatibility also have three components: equality as well as equity in order to ensure cultural compatibility; liberty as well as freedom to enable political compatibility; and solidarity as well as subsidiarity to guarantee economic compatibility.

The first of the collective items, equality, has been a formal constituent of justice since ancient Greece, substantialised as equal respect for human dignity [Sandkühler 1999, 502-503]. The second one, freedom, is today caught between the communitarian and the libertarian extremes, the former attempting to follow the notion of a positive freedom to do good for the sake of the community by choosing that which is necessary to uphold cohesion, and the latter purporting the notion of negative freedom, that is, personal freedom from whatsoever [Sandkühler 1999, 406]. The third collective item, solidarity, is the missing link in the chain of equality and freedom to save inclusion in justice. It is stated that the term might combine Latin *solidus* (as part of a phrase in the context of being indebted) and *sodalis* (meaning brother or similar) into *solidalis* [Sandkühler 1999, 1484].[p]

The agential objective function of self-invention is therefore subdivided accordingly: cultural actors strive for self-fulfilment, the ability to realise their highest aspirations when acting culturally; political actors long for self-determination, the ability to hold their lives in their own hands when acting politically; and economic actors seek self-reliance, the ability to ensure a standard of living when acting economically. These objective functions are imagined as individual moral items comprising:

---

[p] It is unsurprising that the values of the French Revolution, in particular the first two – liberté, egalité, propagated today as building blocks of so-called "liberal" democracies – are highlighted. However, as in the Charter of Fundamental Rights of the European Union, the third one – fraternité – is not forgotten and is referred to under the term solidarity. Moreover, they are put here into another context: their order is different because they are to match the encapsulation of different subsystems as laid down in the previous subchapter (freedom is matched with politics, equality with culture, solidarity with economy); they are also extended to norms and interests; and they all – including freedom as envisioned by the German philosopher Ernst Bloch [Dietschy et al. 2012, 144-161] – are conceptualised as societal items (collective moral items) and not as private personal ones (individual moral items).

recognition by being supplied with cultural commons and being recognised for cultivating commons; empowerment by commons, enabling political decision-making, as well as promoting the implementation of the commons politically; and security due to guaranteed economic commons and, vice versa, by guaranteeing such commons.

Table 2.6.b. Objectified social-informational build-up in work. Societal and agential objective functions and collective and individual moral items in culture, politics and economy.

	good-society structural setting		**good-life agency setting**	
	objective function of the societal self	**collective moral norms, values, interests**	**objective function of agential selves**	**individual moral norms, values, interests**
cultural sub-system	**self-expression** (providing cultural commons)	**equ(al)ity** (cultural compatibility)	**self-fulfilment** (working culturally)	**recognition** (cultivating commons)
political sub-system	**self-government** (providing political commons)	**liberty, freedom** (political compatibility)	**self-determination** (working politically)	**empowerment** (implementing commons)
eco-nomic sub-system	**self-sustenance** (providing ecomomic commons)	**solidarity, subsidiarity** (economic compatibility)	**self-reliance** (working economically)	**security** (guaranteeing commons)

With regard to structure and agency in the eco-social and the techno-eco-social system, the societal objective function can be specified as self-maintenance of the first, and as self-operation of the second system (Table 2.6.c). Self-maintenance describes the faculty to provide the natural commons, self-operation the faculty to provide technological commons. (The second system is termed techno-eco-social because it is a further specification of the eco-social system. Technology is made from nature by humans.) The corresponding collective moral items are called survivability (the ability of the whole to survive by letting the actors survive in a decent manner) and efficacy and efficiency (the ability of the whole to thrive by endowing the actors with technological means to find solutions that improve social life). That concretises the overall social compatibility as environmental and civilisational compatibility, respectively.

The eco-social system actor's objective function is self-preservation by preserving nature to such a degree that nature can continue preserving humans whenever they work. The correspondent moral item is physical well-being. It can be achieved through sustainable commons provided to the actors and through the sustainabilisation of natural commons carried out by the actors. The objective function of technological actors, those using or producing technologies, can be specified as self-actuation. This is because any actuation by an actor – that is, the performance of any social activity that cannot be reduced to mere bodily activity – involves using certain technologies. The respective moral item image is termed tool literacy here, the ability to use tools – prefabricated commons – and, by doing so, the ability to facilitate the production or use of other commons.

Table 2.6.c. Objectified social-informational build-up in work. Societal and agential objective functions and collective and individual moral items in the environment and technosphere.

	good-society structural setting		**good-life agency setting**	
	objective function of the societal self	**collective moral norms, values, interests**	**objective function of agential selves**	**individual moral norms, values, interests**
eco-social system	**self-mainten-ance** (providing natural commons)	**survivability** (environmental compatibility)	**self-preservation** (working ecologically)	**physical well-being** (sustainabilising commons)
techno-(eco)-social system	**self-operation** (providing technological commons)	**efficacy, efficiency** (civilisational compatibility)	**self-actuation** (working technologically)	**tool literacy** (facilitating commons)

2.2.1.5 *Triple contingency*

The creation of social relations in a good society and the creation of individuals that can conduct a good life depend on each other but need not be fully aligned. Social compatibility and individual compatibility are not identical because they reside on different system levels separated by a leap in quality. They can, however, be reconciled by daily efforts of the actors.[q]

Altogether, the moral norms, values and interests set by the actors – for their individual and for their supra-individual objective functions around the provision of the common good, the commons – are coupled like rights and duties, mirroring each other.[r] What actors look upon their unalienable individual rights – the right to dignity, the right to recognition, the right to empowerment, the right to security, the right to physical well-being and the right to tool literacy – as something that is supposed to be granted, respected and even protected by society as society's duties, as duties of the collectivity. At the same time, these very rights – justice/cohesion/ inclusion as social compatibility, equality/equity, liberty/freedom and solidarity/subsidiarity as cultural, political and economic compatibility, and survivability as environmental and efficacy/efficiency as civilisational compatibility – are claims that equally deserve to be treated as rights of society, rights that the actors themselves are obliged to enact as their deemed duty. This points to a complex relationship.

Using the term contingency as introduced by Niklas Luhmann [2001] yields a relationship of triple contingency (Figure 2.11). Social evolution has successive phases of variation, selection and stabilisation.

---

[q] The COVID-19 crisis is in some members of the European Union, and in certain other Western or westernised countries, a striking example of a schism of collective vs. individual values addressing public vs. personal health. Government strategies to counter the pandemic were countervailed by actors who, erroneously and egotistically, prioritised issues of personal health or profit-making prior over minimising casualties by getting vaccinated in order not to infect others. One explanation for this is the life-style colonisation after World War II by the United States, which traditionally cast freedom as rights of defence against the state. It is an irony of history that, after months of hesitating to take action to prevent a fourth wave, the Austrian government has become the first in the EU to step up to its responsibility to ensure public health when it introduced compulsory vaccination. It's all about understanding unity through diversity: as much diversity as possible and as little unity as necessary.

[r] Reading German philosopher Richard David Precht's book on duty [2021], written with a focus on the Corona crisis, helped me formulate my thoughts on the relation of rights and duties.

(1) **Variation** arises through the plurality of individually imagined moral items, individual as well as collective moral values, norms, interests, rights and duties that are accepted and promoted by the actors. That diversity is repeatedly revived and forms varying intersubjective constellations of individual and collective items – that is the phase of intersubjectification.
(2) **Selection** takes place when particular intersubjectified items take hold and satisfy reproductive and/or transformative objective functions of the social system – that is the phase of objectification of the intersubjective diversity, the forming of an objective unity, the societal fixing of individual or collective moral values, norms, interests, rights and duties.
(3) **Stabilisation** of that unity throughout the suprasystem is the third phase. The societally agreed moral items are reinforced by feedback to the intersubjectivity of co-acting networked groups of actors and additionally to the subjectivity of single actors to yield their acceptance of the items – that is the phase of subjectification, of a dominance of the moral items responded by intersubjectivity and subjectivity.

As long as the individuals are treated as the subjects, the society is the object. Shifting the perspective, however, the suprasystem is a subject as well – a supra-individual one – whose elements, the actors, are objects of its interior environment.

This yields a third contingency that adds to a second and a first contingency of social information (to be dealt with in the subsections below) – a triple contingency:

(1) there is no strict determination how single actors will act (as Firsts) with regard to other actors and contribute to an intersubjectification;
(2) there is no strict determination how the network of actors will act (as a Second) with regard to the system and contribute to an objectification;
(3) and there is no strict determination how the system will act (as a Third) with regard to its interior and contribute to a subjectification.

Despite that and because of it, it is the actors themselves who can utilise the field of indeterminacy within the boundaries of determinacy (this is less-than-strict determinism). They are the ones responsible for doing good or doing bad in any realm of social life and for triggering virtuous circles in the direction of the ideal of a good society or vicious circles in the direction of a failing society.

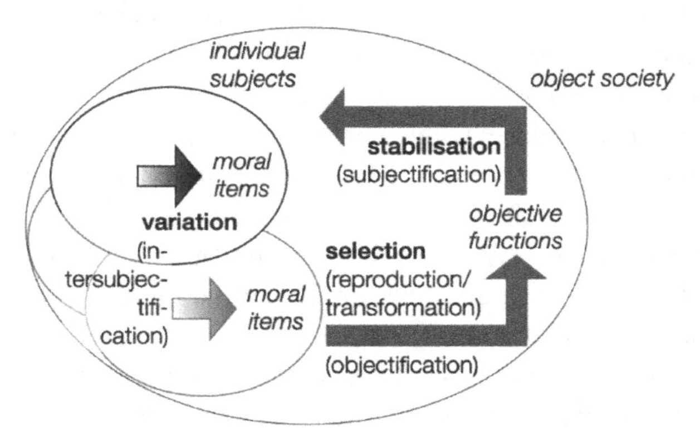

Figure 2.11. Triple contingency of work.

2.2.2 *Remodelling social information: the emergence of discursivity*

Normativity of social information is a constant that is valid for all who produce a communion by setting supra-individual goals. It requires another human asset as a basis – the ability of many actors to achieve consilience. This refers to human communicational social information, which has the characteristics of a discourse. A discourse grounds the dedication to norms and values and interests in the deliberation of the multitude. The deliberation is designed to realise the goals, needing the consilience of many – the popping up of common ideas – to allow collaboration. Here, the term discourse denotes the rational argumentation of possible ways of collaboration such that a solution that satisfies the interests of the many can be found.

How can the existence of such a constant be ensured (Table 2.7)?

The discourse takes the role of the non-being as potentiality, and deliberation takes on the role of the being as actuality.

The individualistic fallacy regards the discursive as a sufficient condition for the ensuing deliberative. This reduction falls short of featuring the full weight of a deliberation that yields consilience.

The sociologistic fallacy falls into the reverse trap of viewing deliberation as a sufficient condition for the discourse, which gives a weight to the latter that it lacks. Discourse alone does not always involve consilience.

Again, the disconnecting fallacy fails to contribute to scientific elucidation.

Table 2.7. The consideration of non-being and being in noogenetic ontology. The multitude's discourse and consilient deliberation.

		non-being	**being**
con-flation	**reduction: individualistic fallacy**	**the discursive:** sufficient condition for the deliberative	**the deliberative:** resultant of the discursive
	projection: sociologistic fallacy	**the discursive:** resultant of the deliberative	**the deliberative:** sufficient condition for the discursive
disconnection: anarchist fallacy		**discursive**	**deliberative**
		independent existents	
combination: noogenetic ontology		**the discursive of the insight into necessity:** necessary condition for a consilient deliberative	**the consilient deliberative:** an emergent from the discursive of the insight into necessity

Noogenetic ontology combines the potential with the actual by considering the discursive social information generation of the multitude as a necessary condition for consilience in the deliberation that emerges from the discourse. Thus, applying Morin's saying about the whole as different from its parts, the consilient deliberative is an emergent. At the same time, it is also more than the discursive (because it adds a new quality) and less than the discursive (because it has less divergence). The consilient deliberation finds a solution to a problem after arguments have been exchanged within the discourse.

2.2.2.1 *Consilience qualities*

The faculty to conduct a discourse leading to consilience reflects the convergence of the multitude's evaluations, interpretations and sensations that pops up through mutual understanding. This faculty is instantiated on different levels of an intersubjectified social-informational build-up in accordance with specific settings of the language. Three levels can be distinguished: the pragmatic one with calls for action, the semantic one harbours arguments and the syntactical one that comprises symbols, whereby the pragmatic is on top and the syntactical at the bottom (Figure 2.12).

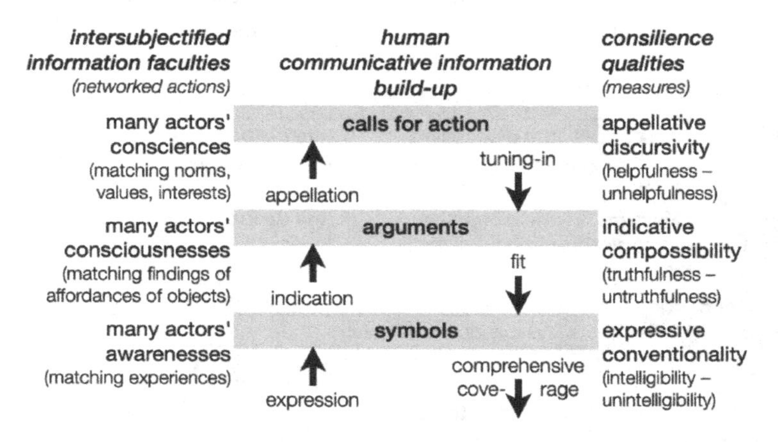

Figure 2.12. Intersubjectified social-informational build-up of language. Faculty instances and language settings for the discourse.

As with consensualisation, the levels are distinctive by representing emergence:

(1) The syntactical feature of social information is achieved by the process of expressing information to a discourse partner; this involves symbols that testify awareness at the lowest quality instance of consilience of communicative social information.
(2) The process of indicating a theme, e.g., by referencing to a certain object of knowledge, contending a certain mechanism that is at work, adds the semantic feature to awareness. This makes the process an issue of consciousness on the intermediate level of the discourse.
(3) The highest level is reached when conscience is implied and calls for action are put forward to the discourse partner by appealing to them. That process is the pragmatic feature in the discourse.

Discourse is a reciprocal process involving at least two actors. When one actor reaches out, the other needs to respond. An appropriate response can take place only if the other can understand the message on all levels, from the pragmatic to the semantic to the syntactic one. Then they can discuss whether or not they agree and whether or not they want to pursue agreements.

(1) **On the pragmatic level**, language is set to let the claims made by the communicator converge with the claims of the recipient. Such an appellative tuning-in of both sides is a precondition for a collaboration in tasks to achieve desired common goals.
 In that vein, each of them can expect messages from the other that are helpful to design and assign those tasks. Helpfulness is thus the most important feature of the noogenetic constant of consilient discursivity.
 It already originated with the dyads of the first step of noogenesis. Messages are sent "for *their* not *our* benefit" and "the communicator highlights for the recipient that he has some relevant information for her" [Tomasello 2014, 52]. The triads merely superposed the dyads' joint goals with collective goals.

(2) **On the semantic level**, language is set to make the meaning of the theme that the communicator ascribes to his message compossible with the interpretation the recipient ascribes to it to yield an indicative fit. This fit is necessary to implement harmonic and meshing actions dealt with by the pragmatic level.
 The recipient needs to trust the communicator to convey to him/her truthfully messages that support and do not undermine the distribution of tasks. The communicator is able to provide messages that fulfil the criterion of considered true. Truthfulness originated with helpfulness. For a message to be helpful it must refer to truth.
 In the dyads, communicators made "a commitment to informing others of things honestly and accurately, that is, truthfully" [Tomasello 2014, 51]. In triads "the individual no longer contrasted her own perspective with that of a specific other – the view from here and there; rather, she contrasted her own perspective with some kind of generic perspective of anyone and everyone about things that were objectively real, true, and right from any perspective whatsoever – a

perspectiveless view from nowhere" [Tomasello 2014, 122]. According to Tomasello, the perspective of Mead's "significant other" that early humans internalised and referenced has been overlaid by the perspective of Mead's "generalised other" with the advent of modern humans [Mead 1934]. Truthfulness of the communicative social information is another feature inherent to the consilient discursivity constant.

(3) **On the syntactic level**, language is set to conventionalise expressions to achieve an expressive coverage of both actors. Without that, the hope for any true or helpful information on the semantic and pragmatic levels were futile.
In order to facilitate that process, convention enables the communicator to pay attention to the comprehensibility of his/her message and the receiver is called to strengthen his/her symbol literacy, which is also possible. Intelligibility is the third feature of consilience characteristic of the noogenetic constant of discursivity.

All uncertainties of the conversation can be questioned and clarified by continuing the conversation until living together becomes possible. Note, however, that expressive conventionality, indicative compossibility or appellative discursivity seem impossible in an absolute sense.

2.2.2.2 *Double contingency*

Discourse ability is contingent as well because at least two actors are involved. At the onset, actor A takes the role of a subject. He has already created an internal model of another actor B. She – B – takes the role of an object for A. The model comprises what A expects from B. A delivers a message to B (Figure 2.13.a).

Subject actor B receives the message, which is a cognitive irritation for her. It contributes to creating a self-organised internal model of actor A and what B can expect from him, who in B's perspective is the object. In line with that model, B sends a message to A (Figure 2.13.b).

The message from B is received by A as a cognitive irritation that self-organises an update of A's model of B and what to expect from her; this can now function as the point of departure for a new message from A to B (Figure 2.13.c).

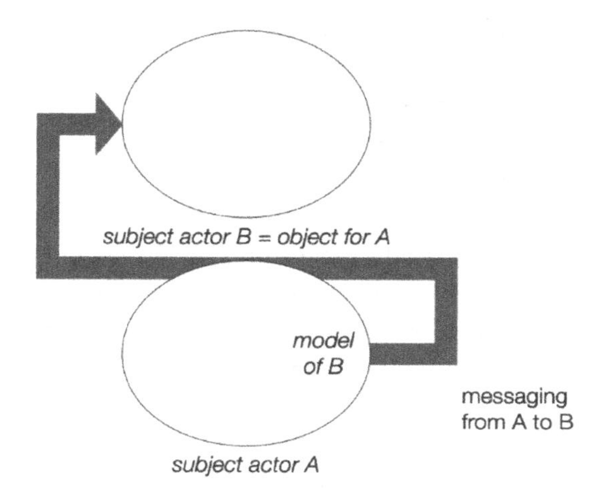

Figure 2.13.a. Double contingency of discourse. A messages B.

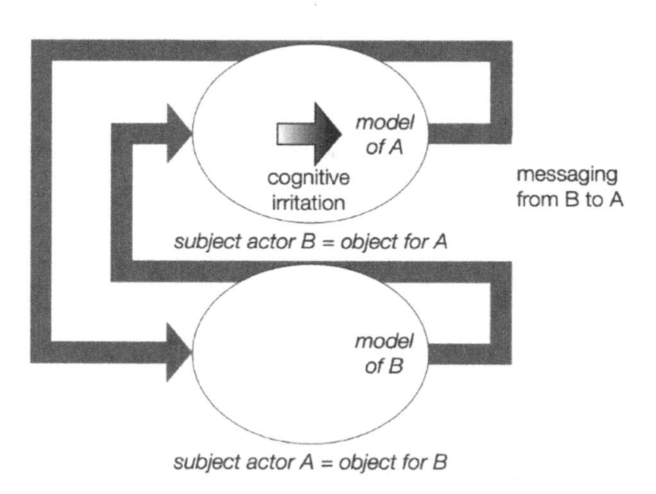

Figure 2.13.b. Double contingency of discourse. B interprets A's message and messages A.

Thereafter, conversation can continue by iterations of the above processes. From actor A's perspective, it is not strictly determined how his reaction to actor B's message will turn out. This is for two reasons: first, it was not strictly determined how her message to him as a reaction to his previous message would be configured and, thus, it was not strictly predictable for him; second, though he now intends to answer her message, it is not strictly determined how to configure this new message – he is free to do so within certain boundaries, since it is an act of his subjective self-organisation. The same holds, *mutatis mutandis*, for actor B. First, she can currently not strictly predict the formulation of actor A's possible new message because it is not strictly determined, and second, she is free to formulate her possible answer within the boundaries of her subjective self-organisation, which makes her possible answer not strictly determined. That is the double contingency in communicative social information.

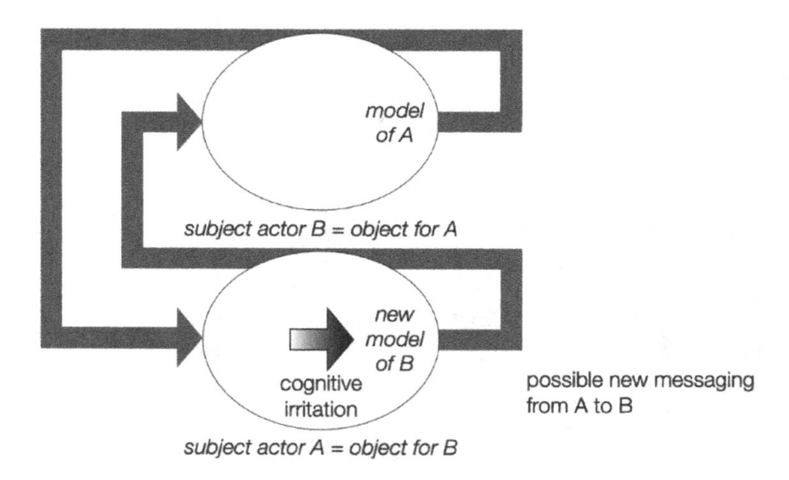

Figure 2.13.c. Double contingency of discourse. A interprets B's message.

Nonetheless, consilience is possible. Each of the three features of the discursive constant – appellative discursivity, indicative compossibility and expressive conventionality – is a Third that can be accessed because self-organisation is taking place.

2.2.3 *Reframing social information: the emergence of reflexivity*

The constant of consensual normativity is bolstered by the constant of consilient discursivity. Equally, the latter is dependent on yet another asset of human creativity: conceptual reflexivity (also a constant).

"Agential subjectivity reflects upon societal objectivity", wrote social realist Margaret S. Archer [2003, 133]. Humans should not be modelled as patients, they are agents. As such they "work reflexively and can monitor their own ways of monitoring society". She upholds the distinction between the model of humans as patients and their model as agents, "[...] between the model of the person as someone to whom things just happen and the model of man as one who seeks to shape the contours of his own life by balancing his enablements and constraints with his concerns" [147]. Whereas "[h]uman *reflection is the action of a subject towards an object*", "[t]he distinguishing feature of reflexivity is that it has the self-referential characteristic of 'bending-back' some thought upon the self, such that it takes the form of *subject-object-subject*" [Archer 2010, 2]. She defines:

> 'reflexivity' is the regular exercise of the mental ability, shared by all normal people, to consider themselves in relation to their (social) contexts and vice versa. [Archer 2007, 4]

Being reflexive, actors consider how to contribute to reproducing and/or transforming the structure of the societal system they live in. They need to discern what they are socially concerned about. This discernment of their social concerns, which is the basis for endorsement and engagement, is a necessary thought input to the deliberation on collaborative tasks in the division of work they are dedicated to. This is because the thinking of an actor must co-ordinate, that is, understand, anticipate and control, which operations he or she will be able and willing to carry out in the context with others.

This calls for putting reflexivity and discernment in perspective with the requirements of the new paradigm. For Archer, the reflexive is something that takes place in internal conversations of an actor – conversations with him-/herself – and can be revealed by empirical research. Nonetheless, the matter that is to be discerned pertains to something relational, which is not so easy for the actor to grasp. From the viewpoint of social critique, it is hidden to every actor and its essence can only become intelligible by efforts of interested people and with the help of critical social theory. How do the reflexive and the discernible fit together (Table 2.8)?

Table 2.8. The consideration of the apparent and the essential in noogenetic epistemology. The reflexion of good life and conceptual discernment.

		apparent	**essential**
con-flation	**reduction: individualistic fallacy**	**the reflexive:** sufficient condition for the discernible	**the discernible:** resultant of the reflexive
	projection: sociologistic fallacy	**the reflexive:** resultant of the discernable	**the discernable:** sufficient condition for the reflexive
disconnection: anarchist fallacy		**reflexive**	**discernable**
		incommensurable knowledge	
combination: noogenetic epistemology		**the good-life reflexive:** necessary condition for a conceptual discernible of the good society	**the conceptual discernible of the good society:** an emergent from the good-life reflexive

When discussing the different approaches to examining the reflexive and the discernible together, the pro and con arguments follow the already known way.

Conflation is set to equalise the apparent and the essential, either by reducing the discernible to the reflexive or by projecting the former to the latter. This involves making one of the sides the resultant of the other side (the latter is then an alleged necessary condition). Levelling down and levelling up are the failures of reductionism and holism.

Disconnection asserts that the apparent and the essential, the reflexive and the discernible, are completely disjunct. This is the failure of anti-hierarchical, separatist thinking.

The only meaningful combination is that the apparent is the necessary condition of the essential, the reflexive the necessary condition of the discernible. This condition does not strictly determine the essential discernible, but renders it an emergent of the apparent reflexive. That is the conclusion of noogenetic epistemology.

2.2.3.1 *Conceptuality qualities*

The individual faculty to be conceptually reflexive, i.e., to produce conceptual reflexive cognitive information, has – compared with the multitude's faculty to conduct consilient discourse – corresponding instances on three levels with special settings of thought (Figure 2.14).

The three levels of subjective social information are, starting from the bottom and following the build-up by emergent leaps,

(1) data, facts and figures,
(2) knowledge, and
(3) wisdom.[s]

In reversed order, the levels nest each other such that the higher level demands the provision of determinate functions from, and by granting relative autonomy to, the lower level as follows [Hofkirchner 2021]:

(1) **Wisdom** is located on the uppermost level. It represents the highest level of human cognition, referring to individual or collective actors. The instance of wisdom is conscience, giving home to norms, values and interests. Wisdom arises through evaluation and judgement of knowledge by a dialectics of processes that also include descriptive features (from below) and prescriptive ones (top down). Evaluations and judgements can be deemed (more or less) right or wrong. But they depend on knowledge from the level below.
(2) **Knowledge** is situated on the intermediary level, representing the mediating linkage between the higher and the lower ends of human cognition. It comprises interpretations of the actor's experiences, instantiated in consciousness. Knowledge arises through concepts, available data, facts and figures, as well as a dialectics of instruction

[s] The well-known sequence of computer science, data – information – knowledge (– wisdom), is somewhat different here. "Information" is skipped because it is the generic term of UTI. The three remaining significations are adapted to match the syntactic, semantic and pragmatic levels of language. Of course, the categories used here can be further divided for deeper granularity.

(from below) and construction (top down). Concepts can be rated (more or less) true or false, independently of the evaluations and judgements of the level above. If true concepts are not coherent with evaluations and judgements, the latter are proven wrong and require correction. Concepts depend on the data, facts and figures of the level below.

(3) **Data, facts and figures** are positioned on the undermost level and represent the most basic level in human cognition. They comprise observations by the actor through sensually experiencing his or her internal and external environment. Those experiences reside in that person's awareness as their proper instance. A dialectics of receiving (from outside) and conceiving (outward) processes is the motor for establishing this kind of social information. Observations can be (more or less) multisided or one-sided, independently of the concepts of the level above. If multisided observations are not coherent with concepts, then the latter are proven false and require overhauling.

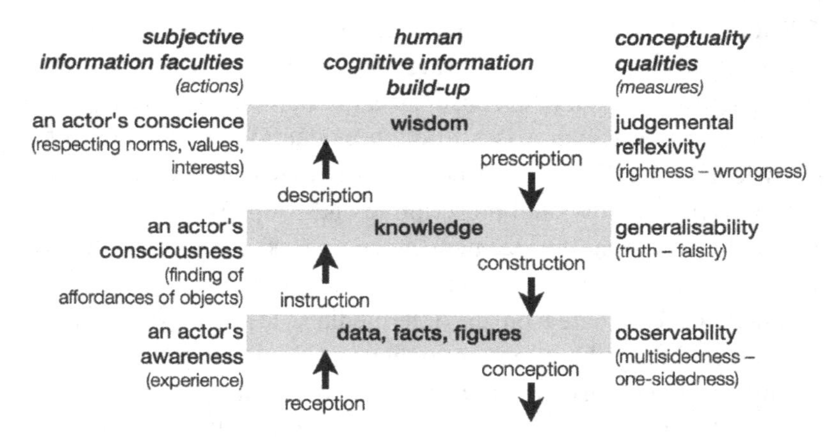

Figure 2.14. Subjectified social-informational build-up of thought. Faculty instances and thought settings for reflexivity.

2.2.3.2 *Concepts*

According to Robert K. Logan [2007], human cognition is distinctive from other living systems' cognitive capabilities through its use of concepts instead of percepts only. As demonstrated in section 2.2.1, the emergence of a new way of co-operation triggered the build-up of social systems and, with it, noogenesis. The objective condition for the origin of conceptuality in human cognition is the build-up of social systems that are hierarchical in that social relations exist on a macro-level that constrain and enable the interaction of actors on a micro-level. Those social relations are the Third that mediates any interaction of the actors as a Second and any action of an actor as a First. There are two more, subjective conditions:

> Subjective condition 1: Actors are able to distance themselves from the system they are elements of. They can reflect on the social macro-level (morals and else) in order to understand the functioning of the social system (its maintenance and its change). They are able to reflect the build-up of social systems. They are able to reflect on the social relations as a Third. This is the origin of social systems thinking. [Hofkirchner 2020b]

The argument is that by reflecting on the circumstances of their inclusion in social systems, nascent humans rudimentarily anticipated the basics of social systems thinking. Otherwise, practice would not have been possible. As Archer demonstrated, actors can go beyond merely reflecting on the reality described by the objective condition. They even can act in various ways vis-à-vis the objective condition because they put what they reflect in the context of their concerns, that is, they are able to create reflexive information. Reflexivity by information they generate is the mediator between their subjective orientations and the objective nature of society. And the better they understand the objective nature of society, the better they can pursue their subjective orientations. This explains why they produce concepts to construe an image of their position in society that is as correct and sophisticated as possible. These concepts have two features, described in systems terms:

> Feature of systems thinking 1: Systems thinking needs to reflect the emergent property of any system, supervenient on the properties of its elements, and not reducible to the latter. Thus, it needs to model emergence in a way that the emergent property is not derivable from premises that describe the properties of elements or their interaction. It has to acknowledge a leap in explaining/understanding according to the leap from a lower to a higher level in reality. It does so by introducing a meta-level in thinking. The level below the meta-level is a necessary condition for the meta-level but not a sufficient one. In that way, the meta-level is itself emerging

> from the lower level. It is the ideational Third that has the task to reflect the Third in reality.

> Feature of systems thinking 2: Systems thinking provides the basis for conceptuality. Concepts [...] are meta-level emergents. They emerge through generalisations. Any generalisation executes a leap from a finite number of phenomena to the class of all possible phenomena that are considered to belong to the same class of phenomena, which, as a rule, represents an infinite number of phenomena. The conclusion from the finite number to the infinite number is not a compelling one. (Only in case the class is set to a finite number, you can execute a complete induction, which, in fact, is a deductive conclusion, since the truth value is transferred from the sum of the single instances to the class.) Concepts are the ideal means for transporting the meaning of systems. They are ideational Thirds. [Hofkirchner 2020b]

The conceptuality of human cognitive social information establishes ideational meta-levels and produces generalisations to cast meta-level concepts that can work as building blocks of specification hierarchies. It does this by creating ideational Thirds. This makes this conceptuality the means to produce information about social life. Importantly, since this works for social life, it can naturally be extended to the world as a whole. The reflexion of the Third is a model for every mode of thinking when confronted with complexity. It is a model for grasping the general relationship between elements and system, parts and whole, of which the individual and society are merely the model instantiation. It is a model for generalisations and subsuming of the specific under the general. This is the second subjective condition of conceptual thinking:

> Subjective condition 2: Actors can use their social systems thinking as template for the understanding of the functioning of any other (non-social) part of the world. The organisational relations on which they reflect is the Third in those systems. This is the origin of systems thinking proper. [Hofkirchner 2020b]

2.2.3.3 *Relational subjects and their relational goods*

Conceptuality renders reflexive actors "Relational Subjects": "A Relational Subject is a subject who exists only in relation and is constituted by the relations that he/she cares for, that is, the subject's concerns" [Donati and Archer 2015, 55]. Relational goods "are goods generated from the *relations between* subjects, ones that remain continuously activity-dependent and concept-dependent upon those involved but cannot be reduced to individual terms" [65]. They "have causal properties and powers that *internally* influence their own makers. They are known [...] by those diachronically responsible for their emergence and synchronically for their reproduction, elaboration, or destruction" [66]. They

> [...] consist of social relations that have a *sui generis* reality; they are produced and enjoyed *together* by those who participate in them; and the good that they entail is an *emergent effect* which redounds to the benefit of participants as well as of those who share in its positive percussions from the outside, without any single subject's having the ability to appropriate it for him/herself. [200]

In line with what has been argued here so far with respect to the commons, the social relations that are ideally set to bring forth relational goods as the common good, as commons, are not antagonistic and not agonistic but synergistic. Nonetheless, relational goods

> [...] can be and, indeed, are competitive goods, but in terms of solidarity in the sense of competition (*cumpetere*) as in the search for the best solutions in a contest which is not detrimental to the other participants but stimulates each participant to contribute his/her best effort toward achieving the same common goal. [219]

2.2.3.4 *Typology of reflexivity*

According to Archer, who elaborated the category of reflexivity, and to Donati [2015], there are several types of reflexive information generation, that is, how the actor can refer to the general level of the social order. In the discernment phase, the individual actor sorts out what matters most to him/her [127-142]. Defining an ultimate concern is the precondition for carrying out the deliberation and dedication phases that lead to the collective commitment. Donati and Archer present an orchestra performance as an example. Relational subjects constitute "a collectivity that evaluates objectives (discernment), deliberates about realizing its common concerns (deliberation), and commits itself to achieving them (dedication)" [61-62].

> If the musician thinks only about himself, he will consult his personal model of perfection and nothing more (in an exercise of Autonomous Reflexivity. If he thinks

> about his own contribution to the orchestra, he will seek how best to 'adapt' his performance to the other players' performance (in an exercise of Communicative Reflexivity). If, instead, he reflects *on* the orchestra's performance and about how this performance could be improved were the musicians to relate to each other in a different way, he will seek to alter the performance of the whole orchestra, that is, he will seek to produce a different emergent effect – a better performance by the orchestra. In this latter case, we can speak of the player practising Meta-reflexivity. [60-61]

Thus, a meta-reflexive individual differs from an autonomous-reflexive individual who disregards the Third because of self-concernedness; he/she also differs from a communicative-reflexive individual who disregards the Third from a standpoint of falling back to what others say. The point is "about *the orientation of all the musicians to the collective performance*. A collective orientation to a collective 'output' is the core of collective reflexivity". The "group is oriented to the relational goods it produces, to maintaining or improving upon them – and to eradicating any relational evils detected in their collective performance" [61].

Reflexivity enables humans to reflect upon themselves as part of a bigger picture, ultimately all the way up to the overarching society itself. The actions of members towards other members of society are mediated by the structure of society, which is a Third in comparison to their own self that is a First and another member that is a Second as long as the First interacts with the other in a direct way and not via the Third [Hofkirchner 2017b, 291-292].

2.2.3.5 *Reflexivity of ego and alter: I – You; Me – Us – Thee*

Reflexivity evolved with the roles the actors played according to how they displayed new cooperation modes and how they conceptualised their own self, the self of the other and the self of all together.

I – You. In dyads, *ego* – the "I" – developed the ability to become conscious of his/her wants when communicating with *alter* – the "You" – about a possible joint venture; both I and You developed the ability to slip ideationally into the role of one another and fancy how the mind of the other might understand the own contribution and the contributions of the other to their common performance (known under the misnomer "theory of mind"); and they developed the ability to evaluate those contributions and to anticipate norms by a sense of fairness that governed their common task co-ordination as a yet situated, unsustainable, ephemeral We. All of this was the germ of a possible social system to be unfolded when the vanishing We was consolidated (Figure 2.15).

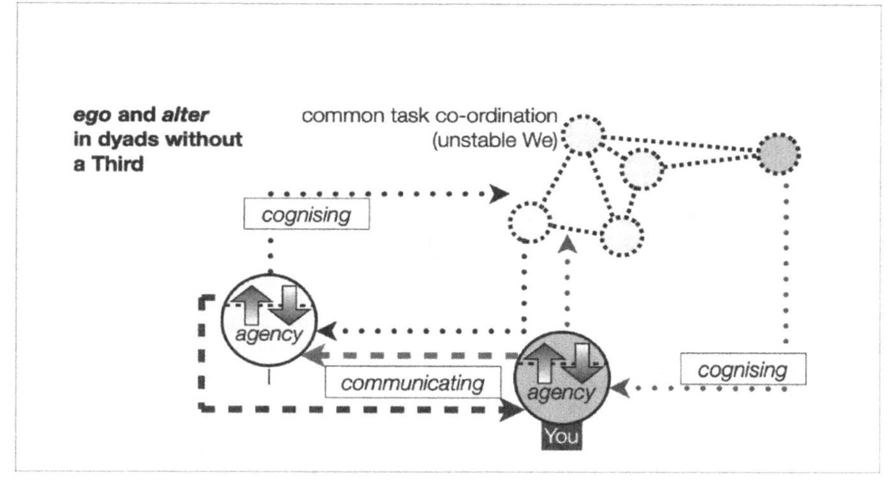

Figure 2.15. Dyadic reflexivity of *ego* and *alter*. I – You.

Me – Us – Thee. In triads, the reflexivity of *ego* and *alter* changed in that it was adapted to suit a higher-order meta-level. They created that level to reflect the objectified social relations that took their place on a macro-level of the emergent social systems – the Third of the commons as the new "Us" (Figure 2.16). This Us – Mead's "generalised other", a conceptual generalisation of the many You's, standing for the whole of society[t] – has become that collectivity that has been mediating since between *ego* and *alter*. *Ego* has become a "Me" that is the I as seen from the perspective of the Us; *alter* has become a "Thee" that is the You from the perspective of the Us; and Me and Thee are related by the Us to each

[t] And not the interpersonal relationship between I and Thou as which Austrian philosopher of Jewish religion Martin Buber's "Ich und Du" [1923] was translated into English.

other such that when Me reaches out to Thee, he or she will do so by factoring in the role that he/she believes is expected from him/-herself by Us. That person will also factor in the role he/she believes is expected from Thee by Us, and he/she will react by the same reflexions, *mutatis mutandis* (see the black arrow from Me to Us and that from Us to Thee, as well as the grey arrows from Thee to Us and then to Me – which illustrate the detour over the Us instead of a direct connection between Me and Thee). Their creation of co-operative information is entangled.

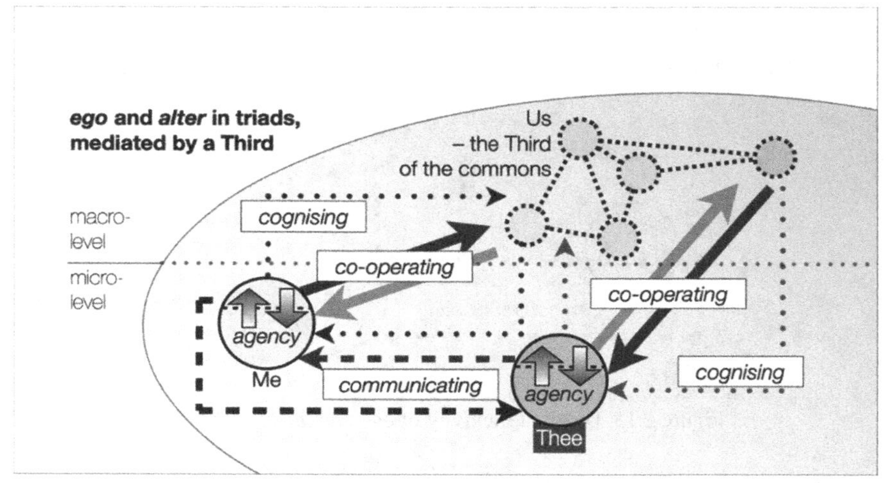

Figure 2.16. Triadic reflexivity of *ego* and *alter*. Me – Us – Thee.

2.2.3.6 *Single contingency*

The figures above sketch any actor as a systemic agent. That agent represents a whole who manifests two levels, a micro- and a macro-level. A dialectics of upward and downward processes occurs between them: the upward process leads to the emergent conceptualisation of qualities on the macro-level, the downward process reworks the micro-level for a better fit with the dominant conceptual qualities of the macro-level. The architecture of single reflexivity is outlined in Figure 2.17.

An intra-systemic nesting of information-generating processes of the actor as informational agent is evident. The horizontal line signifies the borderline whose transgression from below upwards means a shift in quality towards emergence (higher level of integration). A transgression from above means a shift from the dominant quality towards a necessary precondition that can be shaped accordingly (lower level of integration).

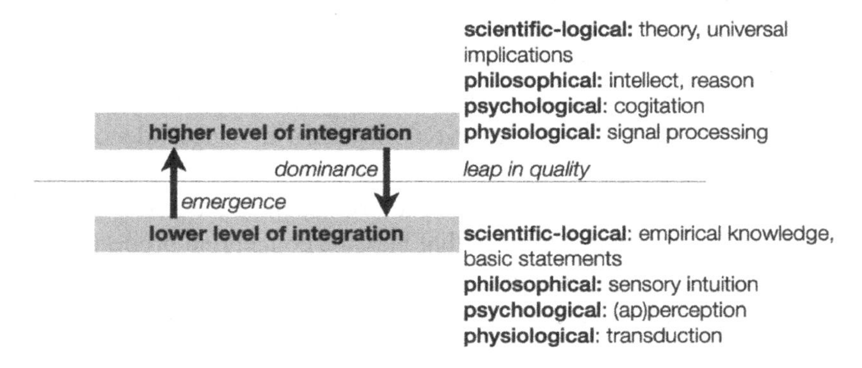

Figure 2.17. Single contingency of reflexion.

First, within certain disciplines, transitions can be discerned between phenomena pertaining to that discipline. Second, transitions can occur between different phenomena that are each dealt with by different disciplines. In pursuing a transdisciplinary approach (obligatory for the new paradigm), transitions of phenomena from one discipline to another must be explored based on which discipline deals with lower-level phenomena and which deals with higher-level phenomena. For the sake of simplicity, both occurrences are exemplified here by a very raw selection of disciplines and phenomena [Hofkirchner 2021, 174-176]:

(1) **Physiology.** The most basic level is the physiological one. Here, in the physiology of cognition, a dialectics of transduction from the outside and signal processing inside takes place. Moreover, the physiological is considered the basis for the psychological level.
(2) **Psychology.** The realm of psychology of cognition features a dialectics of (ap)perception and cogitation that is distinct from physiological transduction and signal processing. Psychology is overruled by philosophy.
(3) **Philosophy.** Philosophically, a dialectics of sensory intuition, on the one hand, and intellect or reason, on the other, has been traditionally

brought forward as distinctive from psychological perception and higher-order psychic functions. General philosophy has been topped by philosophy of science.

(4) **Logic of Science.** From the viewpoint of scientific logic, a dialectics of empirical knowledge and theory exists. In particular, it involves basic statements and universal implications that go beyond the general sensorium and intellect relationship.

In any of the listed dialectics, the higher-level quality of individual cognitive information manifestations signifies a transcendence from a lower-level quality that lacks the conceptual properties that are characteristic of the higher one. In addition, each of the listed disciplines deals with cognitive phenomena that lack the conceptual properties of the phenomena of the next-higher discipline.

The manifold concatenated quality leaps in the generation of reflexive human cognition depart from the activation of neurons and arrive at judgements. This makes single reflexivity contingent because it is not strictly determined how the leaps from the less complex quality to the more complex one will turn out.

2.2.4 *The Principle of Eudaimonism*

Summarising, critical emergentist noogenetics is a cornerstone of a Critical Information Society Theory (CIST). It is an approach that theorises noogenesis, that is, the origin and evolution of social information in social systems. It is based upon the Principle of Commonism and the Principle of the Co-Extension of Information with Self-Organisation. Both together illuminate the co-extension of social information with social self-organisation: as social self-organisation is commons-oriented, social information is commons-oriented, too, and thus critical.

Humanism revisited, update II. Sections 2.2.1 to 2.2.3 argue that there is sufficient reason to assume that the objective setting of emergent social systems as instruments for better co-operation of conspecifics also conditions the subjective setting of nascent and recent humans. That conditioning capacitates creating co-operative, communicative and cognitive social information that directs actors to enjoy flourishing, living a good life in a good society. Aristotle called this state of affairs of individual and society eudaimonia. Eudaimonia is inherent in human unfolding. It is a Third. The British economist and social realist Tony Lawson accounts for that [2017, 239]: "[...] whatever our differences, [...] one factor common to us all is that our flourishing requires that everyone else flourishes. It is impossible for any of us fully to flourish in a system that necessitates that others suffer." Hence "the maxim 'from each according to her or his abilities to each according to her or his needs, in the pursuit of generalised flourishing of our differences'" [240].

Finally, noogenetics can be defined by the Principle of Eudaimonism:

Noogenetics. *Noogenetics is that critical social elaboration of* weltanschauung, *that critical conception of the social world and that critical social-scientific way of creating knowledge that applies the criticist Principle of Eudaimonism, based upon the criticist Principle of Commonism and the informationist Principle of the Co-Extension of Information with Self-Organisation.*

> The **Principle of Eudaimonism** states: there are, regardless of whether or not thematised, associated with the advent of informational social systems in the noogenesis,
>
> (1) the emergence of norms for the plurality participating in a consensualised communion to maintain the system's coherence;
>
> (2) the emergence of discourse ability of the multitude cultivating consilient collaboration to exercise connectivity between the elements of the system;
>
> (3) the emergence of reflexions of the singulars conceptualising their co-ordination to fulfil their elemental responsiveness;
>
> such that normativity, discursivity and reflexivity – thereby specifying the orderly, the corresponding, and the reflective of the informationist approach – are instantiations in a hierarchy of levels. At the same time, each is an emergent meta-level on its own, that is, a Third. These Thirds represent social information of new qualities. They are generated by the social informational agents for different systemic functions from precedent social information, evidencing the creativity of *Homo sapiens*. The aim is a unity of the good society through a diversity of good lives.
>
> Humanism is instantiated as being eudaimonic.

Chapter 3

From Criticism to Critical Utopia

Il y une contradiction indépassable entre les fermetures ethniques, nationalistes, religieuses et le besoin d'une conscience d'humanité commune au XXIème siècle.

[There is an insurmountable contradiction between ethnic, nationalist, religious closures and the need for a consciousness of common humanity in the 21st century.]

– *Edgar Morin, Tweet, Monday, 14 February 2022, 10:00* –

Chapter 2 extended Chapter 1's science of transformation into a science of social transformation and social-informational transformation. The Logic of the Third applies to the social or social-informational real and ideal praxiological, to the potential and actual ontological as well as to the shallow and deep epistemological:

As to the sociogenetic solutions, this logic can consider the common good a Third because the good of the commons re-enters the private as private on loan. It can regard the social as meta- or suprasystemic Third because individual agency is co- or intra-active respectively. It can explore the structural as a Third because the behavioural is ranked as being superficial.

Regarding noogenetic considerations, dedications take the role of a Third when the consensus of the communion emerges from, and dominates the reinforcement of, the sharing of intentions. Deliberation becomes consilient if the discourse is open to what might be labelled concisely an insight into necessity. And the discernment arrives at the concept of a good society if it feeds back to the reflexions on individual good life.

These considerations are all scientific, in particular the theoretical reflections on real logics in a range of subfields of socio- and noogenesis. Such a logic underlies the evolution of social and social informational systems in their totality. Social and social-informational transformation resembles transformation in emergent systems (Figures 3.1.a and 3.1.b):

Systems or proto-elements in a present phase of social evolution are represented by actual social systems or social proto-actors. A space of social possibilities (which is virtual) upgrades the space of overall possibilities. Accordingly, transformation means that, out of several such social possibilities, a structure of a future social meta-system will be selected, whether or not that particular structure is anticipated by the social systems or proto-actors. The leap in quality – the emergence of a higher-order system – they kick off becomes a social revolution, namely the emergence of a higher-order social system. That system is implemented on two levels. On the macro-level the selected structure becomes the actual structure of the new social suprasystem. On the micro-level the new structure turns the social systems or proto-actors of the past phase into actual actors. This justifies speaking of a revolution in the true sense of the word, i.e., whenever social relations are changed to such a degree that they amount to a qualitative leap. A space of new social possibilities is associated with the new, present phase of social evolution.

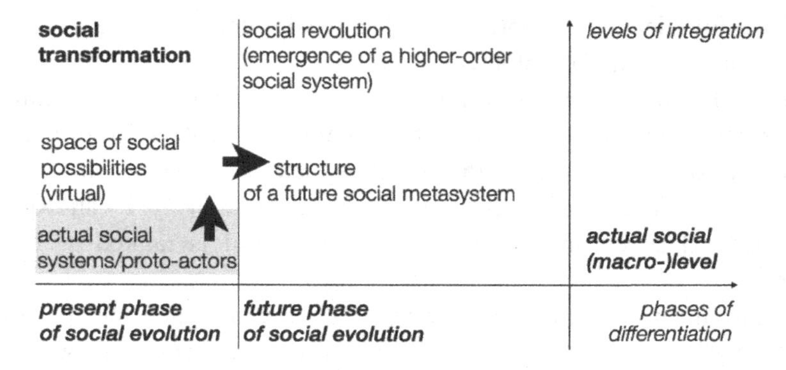

Figure 3.1.a. Transformation in emergent social systems. The case of revolutions – preparatory stage.

This chapter applies the findings of Chapter 2 to the discussion of human evolution when facing the present existential man-made threats. The first subchapter deals with the Great Bifurcation. That phenomenon opens up the objective possibility of a third step in critical sociogenesis –

a step whose actualisation is the Great Transformation. That, in turn, necessitates scrutinising subjective noogenetic factors in the second subchapter.

social transformation	social revolution (emergence of a higher-order social system) space of social possibilities (virtual)	*levels of integration*
	actual structure of the social suprasystem	***actual social macro-level***
past social systems/ past proto-actors	actual actors of the new social suprasystem	***actual social micro-level***
past phase of social evolution	***present phase of social evolution***	*phases of differentiation*

Figure 3.1.b. Transformation in emergent social systems. The case of revolutions – implementation stage.

3.1 Rethinking the Social in the Age of the Great Bifurcation: Emergentist Bifurcationism

This subchapter resumes the Critical Social Systems Theory (CSST) cornerstone of critical emergentist sociogenetics. The main idea of the latter is that the existence of commons is essential for humans and that humanism is essentially commonist. That idea is key to understand the crises sociogenesis is undergoing today. The root cause of the critical state humanity finds itself in is the improper treatment of the commons.

> Karl Marx focused on the dialectic of productive forces and relations of production. Critical social systems thinking in the age of global challenges needs to extend that focus on a dialectic of, on the one hand, social actors that are more and more concerned about the common fate of humanity in the age of global challenges and, on the other, social relations that, on every level of society, are, in principle, relations of commoning though having been turned more and more into antagonistic relations. [Hofkirchner 2017b, 286]

The commons have become a planetary issue. How they are tackled will decide on the survival or death of humankind. The trajectory of social systems evolution is on the point of reaching a bifurcation between existence and extermination – the Great Bifurcation.

Bifurcations are ubiquitous. Any complex system experiences bifurcations – conjunctions at which the objective trajectory that the system is following splits up into at least two possible trajectories. The system must take and continue on one of the different paths. Depending on the granularity of the system theoretical analysis, the difference between the paths might have a minor or major impact on the system. Small-impact paths are taken more often than large-impact paths. The next bifurcation is then a different one.

The term level-evolution is applied as long as systems follow a trajectory that does not alter the steady state, maintaining some sort of equilibrium [Haefner 1992; Oeser 1992] (Fig. 3.2.a). This means that no qualitative change occurs even though the system might cross a series of bifurcations with negligible changes. The path taken shifts the varying space of possible trajectories from one instant to the next. Evolution continues. Dotted lines in figure 3.2 symbolise possible trajectories.

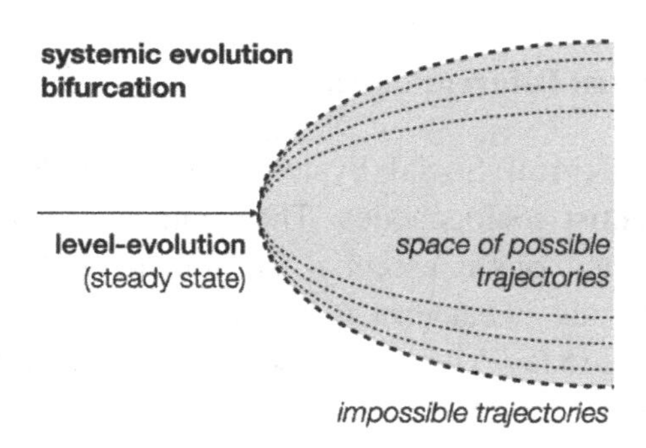

Figure 3.2.a. Bifurcations. Level-evolution.

Alternatively, approaching bifurcations with large impact trajectories involving qualitative changes leads to amplified fluctuations in the current trajectory. The system is in crisis, evolution is punctuated – a term borrowed from the palaeontologist Stephen Jay Gould, who cast the picture of punctuated equilibrium in the course of biological speciation [Gould 2002] (Figure 3.2.b). The crisis is mastered when the system breaks through the punctuation to a path that successfully raises the complexity of the system. This path is called mega-evolution [Haefner 1992; Oeser 1992]. Mega-evolution depicts the whole line from a beginning through all bifurcations that have brought qualitative change to the system. Another trajectory, however, can also lead to a decline in complexity, a breakdown of the system and devolution into lower-level constituent parts of a bygone history.

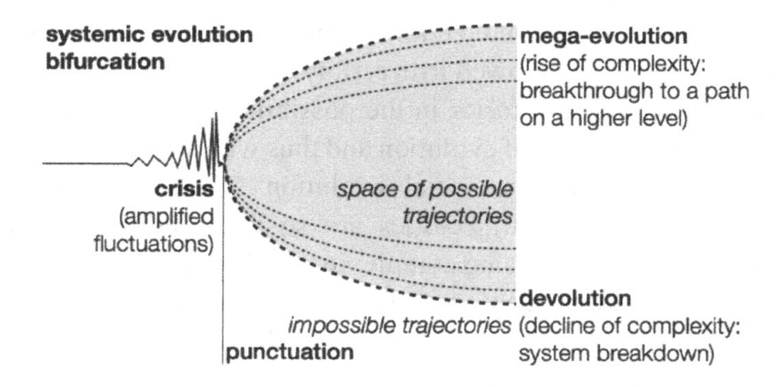

Figure 3.2.b. Bifurcations. Mega-evolution.

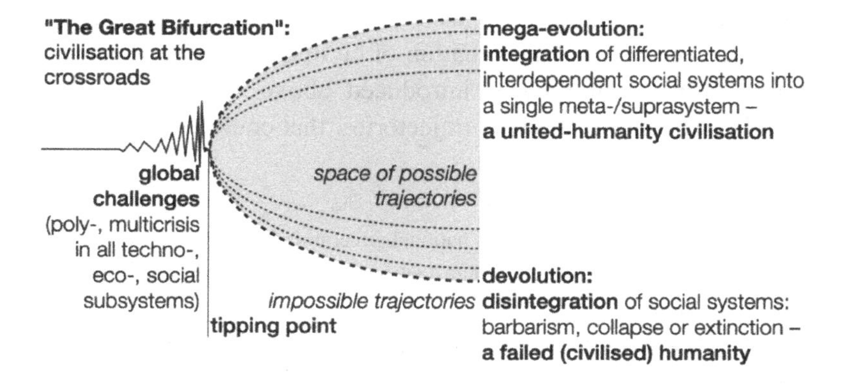

Figure 3.2.c. Bifurcations. The Great Bifurcation – civilisation at the crossroads.

The same bifurcational pathway holds for sociogenesis (Figure 3.2.c). Global challenges cause increasing fluctuations, featuring a poly- [Morin 1999] or multicrisis [Brand and Wissen 2021] throughout all social systems. This includes the techno-, eco- and the economic, political and cultural subsystems. The crisis in one field of a social subsystem nourishes the crisis in another field of the same social subsystem, the crisis in one social subsystem amplifies the crisis in another social subsystem, the crisis in one society made up of subsystems boosts the crisis in another society. Earth civilisation as a whole is in a crisis. The tipping point – systems philosopher Ervin László, who never tired of popularising this complex systems insight into the fate of the social world, terms it chaos point [Laszlo 2001; 2010] – is a point of no return. A tipping point, once crossed, means the new trajectory is followed irreversibly.[a]

The number of future trajectories in the possibility spaces have been increasing with the complexity of evolution and thus with social evolution. According to László, however, social evolution faces the strongest bifurcation ever, because the trajectories are so polarised: the mega-evolution that is currently increasing complexity would require integrating the differentiated, interdependent social systems into a single meta-/suprasystem. The task is nothing less than a united humanity. Moreover, any devolution that reduces complexity would trigger the disintegration of social systems, causing an ongoing failure via an unrecoverable dystopia or an unrecoverable collapse, if not even extinction. Those trajectories would amount to an existential catastrophe [Ord 2021, 37] leaving a failed humanity in their wake. This situation of civilisation at the crossroads[b] is the Great Bifurcation[c] briefly introduced above. There may well be intermediate ways, intermediate trajectories, that could be selected. Were

---

[a] Any pandemic is an example. The less national governments take precautions against the exponential spread of the virus, the more incidences, fatalities and virus mutations the world population must suffer.

[b] "Civilisation at the crossroads" was the title I coined for the European Meeting on Cybernetics and Systems Research 2014 in Vienna.

[c] The term appears in a lecture I held in 2004 in Cork. I have to confess that until the writing of this manuscript I was not aware of the fact that Ervin László had used the term well before [1986, 1990]. With choosing that name I wanted to associate with Karl Polanyi's term of the Great Transformation which expounds the problems of reconciliation of democracy with capitalist markets. Notwithstanding, it is my conviction that, though being connected with a task like that, the task of saving humanity is of yet another order.

this to be done, however, the probability is high that they would eventually turn out to bifurcate into the two main trajectories at some later point.

Planetary bifurcationism is the topical social perspective, the topical conception of the social world and the topical social-scientific way of thinking in the age of global challenges. Accordingly, it will be used in the following sections to discuss redesigning, remodelling and reframing of the social (Figure 3.3). The current complex global social systems challenges require the re-design of humanity as a common-destiny community, a new model of civilisation in the so-called Anthropocene, and the new frame of anthropo-relationality.

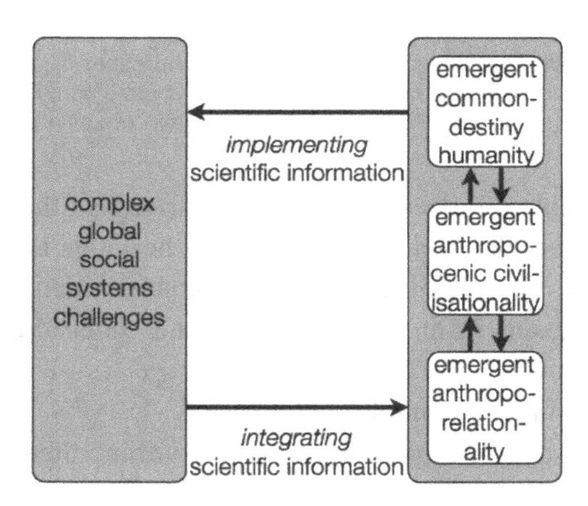

Figure 3.3. Critical emergentist-systemist bifurcationism. The emergent planetary content of the new paradigm.

3.1.1 *Redesigning the social in the age of the Great Bifurcation: the emergence of common-destiny humanity*

Today, there is an "Is" that describes differences among humanity on a global scale that reaches down to a local scale – hence the term glocal [e.g., Robertson 1992] to describe the objective range. The "Ought" is often claimed as a unity knit by universal human values – a likewise objective global humane. Its substantiation is a matter of debate (Table 3.1).

Table 3.1. The consideration of Is and Ought in bifurcationist praxiology. The actors' (affected) glocal differential and the global/common-destiny humane.

		is	**ought**
con-flation	**reduction: individualistic fallacy**	**the glocal differential:** sufficient condition for the global humane	**the global humane:** resultant of the glocal differential
	projection: sociologistic fallacy	**the glocal differential:** resultant of the global humane	**the global humane:** sufficient condition for the glocal differential
disconnection: anarchist fallacy		**glocal differential**	**global humane**
		disparate takes	
combination: bifurcationist praxiology at the Great Bifurcation		**the affected differential:** necessary condition for a common-destiny humane	**the common-destiny humane:** an emergent from the affected differential

The individualistic fallacy cannot demonstrate that the global humane can result from the glocal differential, since the latter is not a sufficient condition for that. The sociologistic fallacy cannot evidence that the glocal differential can result from the global humane, since the latter is not a sufficient condition either.

The anarchist fallacy cannot relate the two.

Bifurcationist praxiology, however, can combine them. The universal humane is no longer a mere cosmopolitan dream. It has become real because social systems have conquered the entire terrestrial surface of the globe. This globalisation process entails an objective interdependence. Actions can have far-reaching and long-term impacts. In fact, "effects external to one society turn out to become internal to other societies. The environment of a society is made up of all other societies. [...] The principles of societal development that have been effective so far, cannot be any longer effective without resulting in serious disadvantages to the maintenance of society" [Hofkirchner 2014a, 67-68]. These problems are known under the label of the global challenges. The universal humane must now be understood as an indication that all humanity has become a community of common destiny. Then, as common-destiny humane, it can be understood as emerging from the global differential that needs to be recast as every affected difference. People are affected by the enclosure of the commons, whether they belong to the tribe of people who have been impeded or even to the tribe that has benefitted so far. Now that externalised impacts of the enclosure are falling back to those who benefitted, all are affected. Importantly, not only thriving is at stake, but in fact the very survival of all human beings and nature as such.

3.1.1.1 *Enclosure of the commons carried to extremes*

The course of history reveals an increase in the enclosure of the commons. To date, this development has reached an extent that constricts flourishing in every system integrated with society. "Self-organisation of the good life has become exclusive, commoning relations have become relations of exclusion that deprive increasing numbers of actors of the commons such that actors have been set in competition with each other" [Hofkirchner 2017b, 285].

The description here departs from the cycles for the good and the bad of section 2.1.1 and follows the architecture of society as nested systems (section 2.1.2).

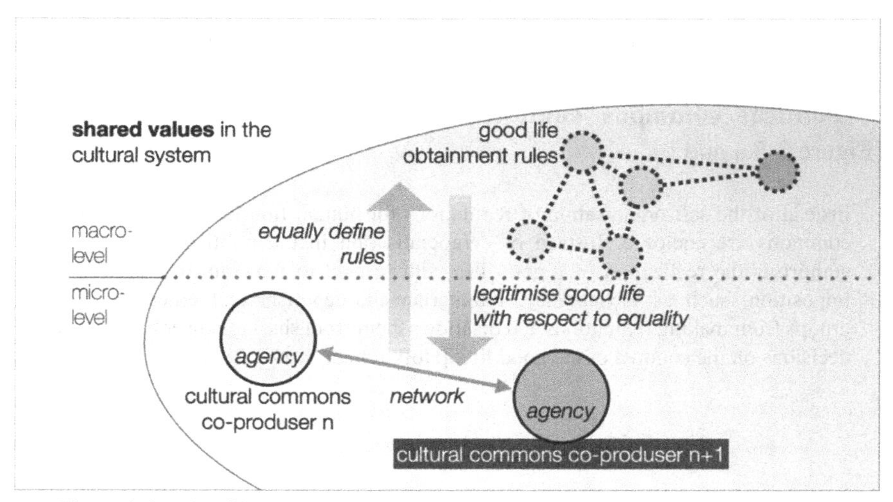

Figure 3.4.a. Conflictive rule definitions. Self-organisation of good life rules.

Cultural commons contested. Hegemony has taken over shared values (Figure 3.4.a and b):

> Instead of the self-organisation of rules for the good life, the cultural commons are enclosed. Instead of sharing values in the cultural system that legitimise a good life with respect to equality, there is rather the hegemony of provincialism, nationalism, racism, anti-islamism, fundamentalism among others that exclude groups of certain other cultural actors from the egalitarian setting of rules and denies them the sharing of power over definitions for obtaining the good life. [Hofkirchner 2017b, 285]

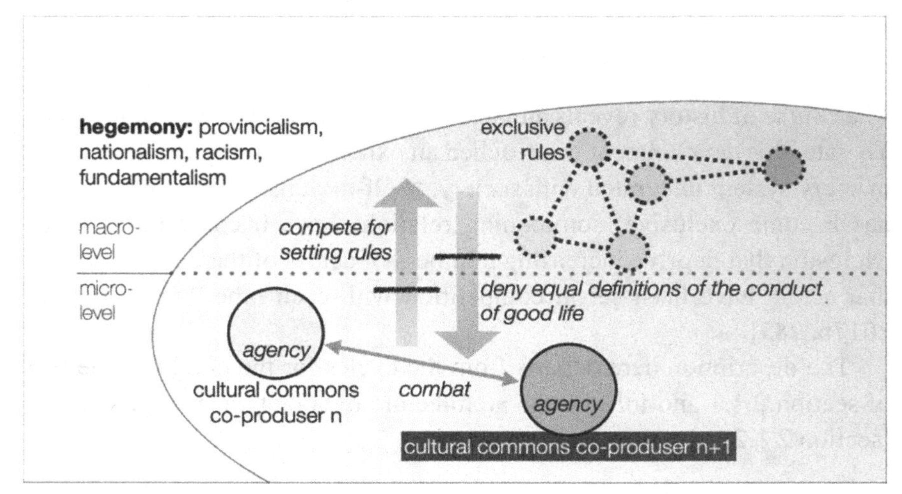

Figure 3.4.b. Conflictive rule definitions. Enclosure of the cultural commons.

Political commons contested. Strength is trumping democracy (Figure 3.5.a and b):

> Instead of the self-organisation of regulations for human flourishing, the political commons are enclosed. Instead of democratisation in the political system that authorises the realization of a good life with respect to freedom, the politics of imposition, such as technocracy, totalitarianism, nepotism and others exclude groups from making regulations and disallows them from sharing power to take free decisions on the conduct of the good life. [Hofkirchner 2017b, 285]

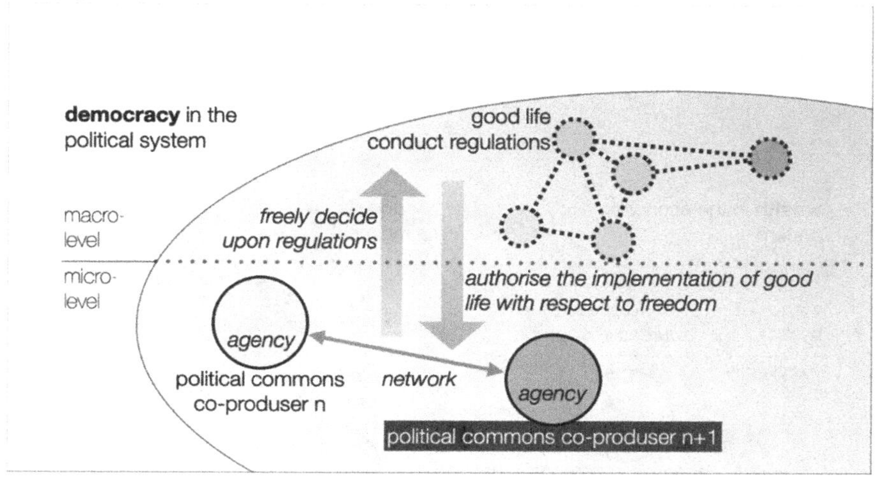

Figure 3.5.a. Conflictive regularity decisions. Self-organisation of good life regulations.

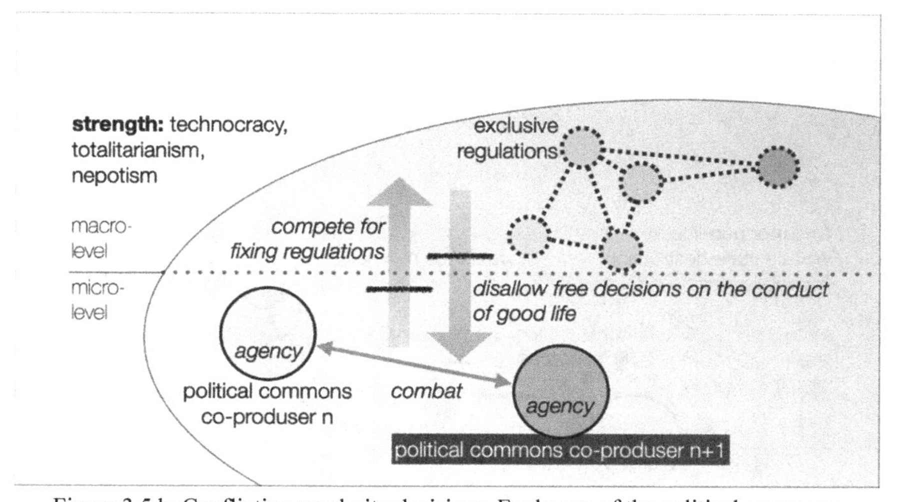

Figure 3.5.b. Conflictive regularity decisions. Enclosure of the political commons.

Economic commons contested. Wealth is replaced by fortune (Figure 3.6.a and b):

> Instead of the self-organisation of resources for the good life, the economic commons are enclosed. Instead of sharing wealth in the economic system that allocates the means for a good life with respect to solidarity, wealth dominates as in

the financialization of capitalism and the neoliberal destruction of the welfare state, excluding some groups from resources and depriving them from any power to dispose over the means for living a good life. [Hofkirchner 2017b, 285]

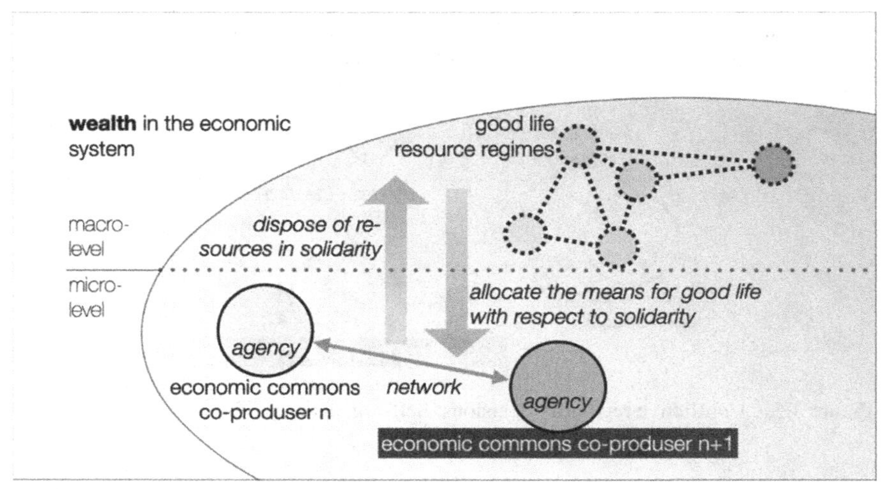

Figure 3.6.a. Conflictive resources disposal. Self-organisation of good life resources.

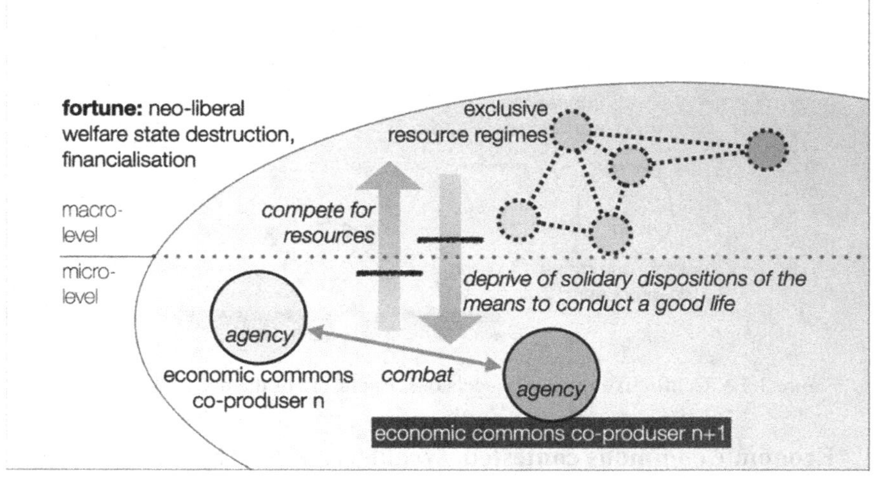

Figure 3.6.b. Conflictive resources disposal. Enclosure of the economic commons.

Ecological commons contested. Colonisation of nature resets harmony (Figure 3.7.a and b):

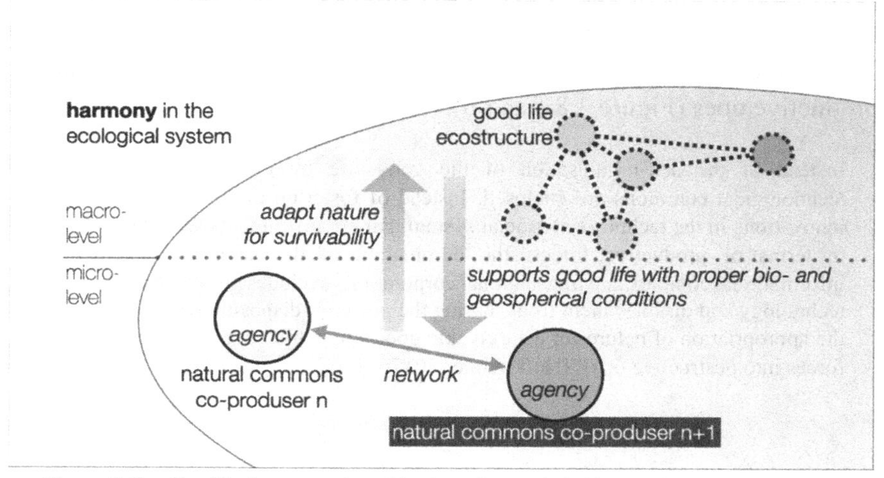

Figure 3.7.a. Conflictive *umwelt* and body safeguard. Self-organisation of good life ecologies.

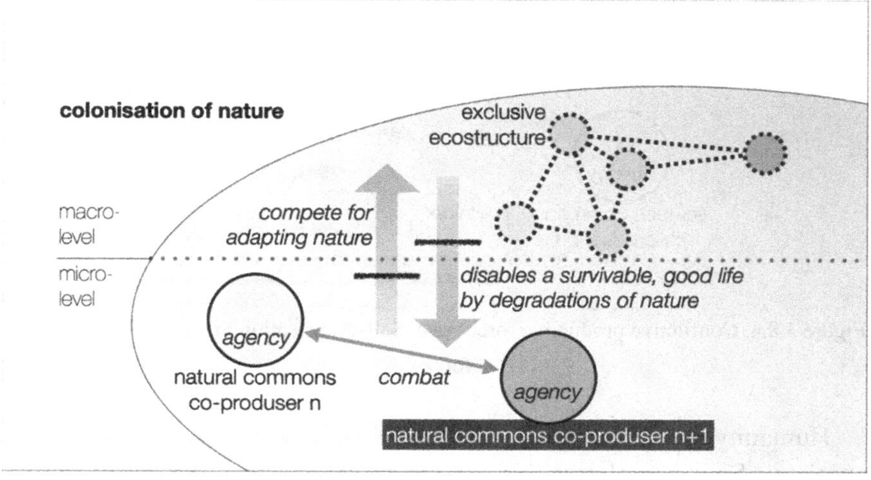

Figure 3.7.b. Conflictive *umwelt* and body safeguard. Enclosure of the ecological commons.

Instead of the self-organisation of ecologies promoting the good life, the ecological commons are enclosed. Instead of harmony in the eco-social system that supports

> the good life with proper bio- and geospherical conditions, nature is colonised, which excludes groups from adapting nature and disables them from sharing the power to appropriate nature for survival, which also leads to degradations of nature. [Hofkirchner 2017b, 285]

Technological commons contested Destructive forces outbalance productive ones (Figure 3.8.a and b):

> Instead of the self-organisation of the good life by productive forces, the technological commons are enclosed. Instead of fostering the meaningfulness of innovations in the techno-(eco-)social system that are instruments for the good life in terms of productive forces, the domination of the latter by a military-informational complex and transnational corporations excludes groups from shaping technology and disables them from sharing the power of disposition over means for the appropriation of nature for a survivable good life, which even turns productive forces into destructive ones. [Hofkirchner 2017b, 286]

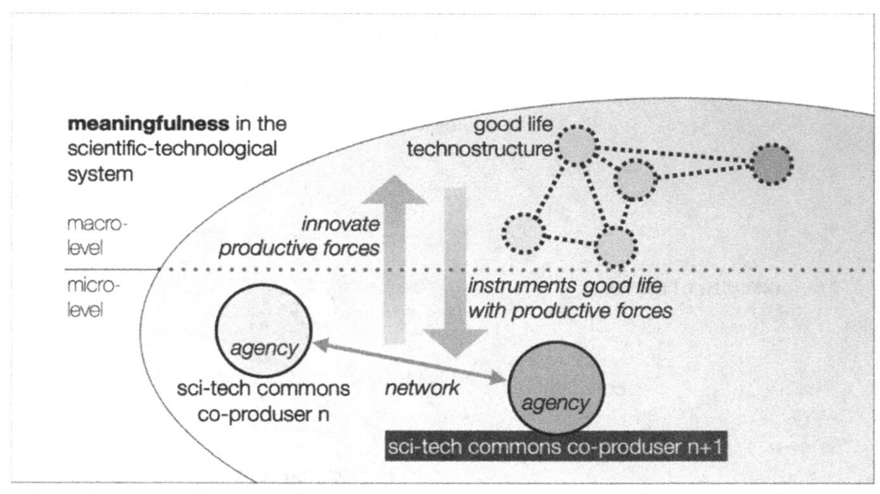

Figure 3.8.a. Conflictive productive forces rise. Self-organisation of good life productive forces.

Humanity as a whole does not yet exist. However, it is a utopia in *statu nascendi*. Social evolution has so far paved the way for an objective community. But it requires a struggle to fight to halt the spiralling-down cycles and "to make the social systems inclusive through the disclosing of the enclosed commons" [Hofkirchner 2017b, 286] in order to warrant eudaimonia, a good life for all in a good society. It requires a transformation of societies into a state that saves the human world from the threat of decline and enables it to survive by making it thrive.

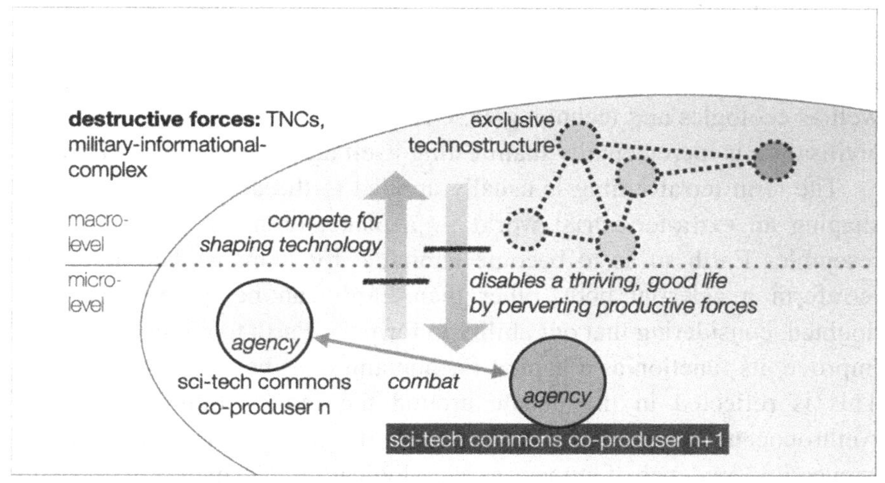

Figure 3.8.b. Conflictive productive forces rise. Enclosure of the technological commons.

Such struggles are currently underway throughout the societal edifice from the technological to the cultural system [Hofkirchner 2014a, 70-71]:

(1) On the science and technology battlefield, there is a struggle for science as a "communist", universal, disinterested and organised sceptical endeavour, as Robert K. Merton put it in 1942 [1973, 267-278], and a struggle for technology assessment, human-centred design and value-based engineering.
(2) On the battlefield of the external and internal nature of humans, there is a struggle for a cautious treatment of the bio-physical basis of human life.
(3) On the resources battlefield, there is a struggle for unalienating working conditions and a fair share for all.
(4) On the agora battlefield, there is a struggle for participatory democracy.
(5) On the battlefield of the community of values, there is a struggle for inclusive definitions of selves.

3.1.2 *Remodelling the social in the age of the Great Bifurcation: the emergence of anthropocenic civilisationality*

Agency and structure are currently undergoing a planetary change. Agency is not restricted locally or regionally; its impact is on a worldwide structure that is a network of different cultures, polities and economies as well as ecologies and technologies. Combined, that network of planetary civilisation is incrementally manifesting itself and shaping planet Earth.

The term terraforming is usually applied to the deliberate process of shaping an extra-terrestrial world – a planet or moon – such that it resembles Earth so as to become habitable for humans. The ability to terraform a celestial body other than Earth can be quite reasonably doubted, considering that our ability to form the Earth to retain, much less improve, its function as a habitat for humanity has become questionable. This is reflected in the debate around the introduction of the term Anthropocene as a geological category to denote the era featuring irreversible evidence of humans as a geological force, which is in line with Vernadsky's intentions and writings. The Anthropocene reveals both our success and our failures. In the present context, terraforming is used for the process of shaping Earth and is afflicted with the same ambiguity as is the term Anthropocene.

Even the term civilisation – generally implying the epitome of human achievement and aspiration and synonymous with a "good society" – has a pejorative connotation. Morin writes about "The civilizing of civilization" as a task for humanity. He concedes that Western civilisation needs to be reformed. But he also concedes: "There is nothing harder to realize than the wish for a better civilization. The dream of personal development for all, of real fellowship between all, of general happiness, has led its champions to the use of barbarous means that have ruined their civilizing enterprise" [Morin 1999, 89]. Thus, civilisation must be civilised, without expecting to be able to realise an ideal civilisation or some absolute utopia.[d]

How can terraforming and civilising be related, given the global challenges, and what is the role of the Anthropocene (Table 3.2)?

Table 3.2. The consideration of non-being and being in bifurcationist ontology. The (anthropocenic) glocal terra-formative and the global/anthropocenic civilising.

		non-being	**being**
con-flation	**reduction: individualistic fallacy**	**the glocal terraformative:** sufficient condition for the global civilising	**the global civilising:** resultant of the glocal terraformative
	projection: sociologistic fallacy	**the glocal terraformative:** resultant of the global civilising	**the global civilising:** sufficient condition for the glocal terraformative
disconnection: anarchist fallacy		**glocal terraformative**	**global civilising**
		independent (potential and actual) existents	
combination: bifurcationist ontology at the Great Bifurcation		**the anthropocenic terraformative:** necessary condition for an anthropocenic civilising	**the anthropocenic civilising:** an emergent from the anthropocenic terraformative

Firstly, it must be stated that actions of local and regional terraforming as such do not civilise the social relations in a way that would rework existing civilisations and produce a global civilisation deserving of the name. The result today is a worldwide aggregation of small-scale enterprises that are not ordered in a longed-for manner. That approach reflects reductionist thinking. It must also be said that a conflation in the opposite direction does not work. A global civilisation as such does not imply a certain way of terraformation. Hence, projectionist thinking is no solution either.

[d] The argument of civilising civilisation reproduces the more abstract argument that there is no other instance, other than reason itself, to criticise reason, in particular instrumental reason in the wake of Enlightenment and even earlier social periods.

Secondly, conceptualising terraformation and global civilisation as completely disjoint from each other foregoes the option of any comprehensible mutual influence on each other.

Thirdly, ontology needs to take the Great Bifurcation seriously and exercise the combination of the potential and the actual. The concepts of terraformation and global civilisation must be cognisant of the anthropocenic dimension of civilising Earth. This dimension is anticipated by Vernadsky's assessment of scientific thought as a planetary phenomenon boosting the transition from the biosphere to an anthropo- or sociosphere in the quality of a noosphere from the outset. With the advent of global challenges as existential risks, this dimension must be reconsidered. Terraformation must signify that the possible actions it consists of have anthropocenic consequences and thus contain that which can be civilised. Global civilisation must express its realisation of the anthropocenic civilisable by means of terraforming in an anthropocenic sense. This approach reconciles terraformation and global civilisation – if the anthropocenic terraformative is the necessary condition for the global civilising that is, in turn, conceived of as an anthropocenic civilisation that takes the role of an emergent.

3.1.2.1 *Anthropocene*

What we refer to here as the Anthropocene must not stick to disciplinary boundaries. It needs to go beyond incorporating geological, chemical, biological issues and other scientific ones that focus on nature to, first and foremost, treat historical and social scientific aspects. This is the only strategy that can yield the transdisciplinary view compatible with the new paradigm represented here. Finally, it requires a decision on where to locate the "read thread" that knits together all the relevant disciplinary issues. If humankind is said to have become a geological force, then humankind's history must form the starting point.

The contemporary time that humanity has entered is an age of global challenges. These global challenges are the evidence that humanity has developed global power. The power that humanity has developed is so strong that it has begun to have repercussions on humanity itself and pose existential risks. These existential risks are evident in three areas:

(1) **Arms race in the "Atomic Age".** The Bulletin of the Atomic Scientists has long been publicising a Doomsday Clock, and every new year they make a statement that resets that clock. The announcement of January 2022 is: for the third time in a row, it is 100 seconds to midnight, that is, the clock remains "the closest it has ever been to civilization-ending apocalypse" [Bulletin of the Atomic Scientists].
Their start point was the atomic bombing of Hiroshima and Nagasaki on 6 and 9 August 1945. Since then, the nuclear weaponry has been – wilfully or unwilfully – threatening the peaceful coexistence of nations. Arms control has somewhat civilised the arms race, but has been unsuccessful in removing all atomic, biological and chemical weapons of mass destruction. In particular, nuclear weaponry control measures involving the US and Russia have not prevailed and have failed to prevent modernisation on both sides.
It is somewhat of a miracle that neither insane intents nor unintended happenstances have so far led to the suicide of the human species (and killed off many other life forms on Earth too). Some close calls have been recorded. No tipping points have yet been reached. However, nuclear fall-out of earlier tests will be traceable for geological timescales. The military – and the constant wars being waged – are the world's biggest CO_2 emitter.
The Atomic Age of weapons is a hybrid: it is also an Atomic Age of the so-called peaceful use of atomic energy – a contested field. Apart from countless hazardous incidents and events that even go beyond the "Maximum Credible Accident", the long-time storage of nuclear waste remains an unsolved problem.
Civil society has responded with peace movements. In 1955, the so-called Russell-Einstein Manifesto, signed by world-famous scientists Bertrand Russell and Albert Einstein, laid the foundation for the Pugwash Conferences and other initiatives.

(2) **Extractivism and industrialism overstretching planetary boundaries.** Sometime later, US biologist Rachel Carson [1962] published her book Silent Spring on the impact of herbicides and pesticides such as DDT on mammals and birds through contamination of water and soil along the food chain. She dedicated

her book to Albert Schweitzer, who had warned against nuclear weapons. Her book launched the environmental movement worldwide.

Today, the moniker "Climate Change" for the denomination of man-made heating of planet Earth has become priority number one in ecology, even though the issue has been recognised since the 1970s. Science has so far identified more than a dozen tipping elements, five of which possibly already switched within the Paris range of 2°C temperature anomaly: the melting of the West Antarctic ice sheet, the Greenland ice sheet, the Arctic summer sea-ice[e], and alpine glaciers as well as the die-off of coral reefs. The other identified tipping elements in the Earth system would be a switch from the Amazon rainforest to savanna, a boreal forest shift through wildfires and pests, slowdown of the thermohaline ocean circulation, perpetuation of the Sahel drought region due to a destabilisation of the West African monsoon, destabilisation of the combination of El Niño and the Southern Oscillation, melting of the East Antarctic ice sheet, thawing permafrost, and loss of Arctic winter sea-ice. All define critical thresholds that, once breached, have non-linear impacts. They can intertwine with each other and with tipping points in other realms in- and outside ecology. This makes it unforeseeable when the temperature rise will come to a halt, which of the initial impacts will cascade into a chain of impacts, and whether or not they will reinforce simultaneous changes [Schellnhuber et al. 2016]. They can potentially add up to one global tipping point [Schellnhuber et al. 2019]. This casts strong doubts on whether lasting increase in global warming can be ruled out.

Global warming is only one of more than a dozen ecological hazards that are of anthropogenic origin. It suffices to point to what is being referred to the 6th major mass extinction.

Of course, the COVID-19 pandemic deserves mention in this context.

(3) **A "Third World War" pitting the rich against the poor.** The social world is extremely unequal: "a tiny group of over 2,000 billionaires

---

[e] Whether the Arctic summer ice represents a tipping element is being contested, as IPCC author Jochem Marotzke reports in an interview [2022].

had more wealth than they could spend in a thousand lifetimes"; "for 40 years, the richest 1 % have earned more than double the income of the bottom half of the global population" and "have consumed twice as much carbon as the bottom 50 % for the last quarter of a century" [OXFAM International 2021]. Well-known scientist and activist Vandana Shiva [2019] maintains that the global corporations of one per cent of the world population cause poverty, malnutrition, and a refugee crisis. French economist Thomas Piketty [2014] theorises that wealth and income inequality in Europe and the United States tends to enlarge. He shows that, historically, relief went hand in hand with wars. Political scientists Ulrich Brand and Markus Wissen show in their book that one part of humanity is producing and living at the cost of another part and thus at the cost of natural living conditions. They term this an "imperial mode" of production and living [Brand and Wissen 2021]. In addition, social criticist Noam Chomsky [2021, 90-120] warns in a lengthy interview – exclusively available in the German edition of his book "Internationalism or extinction" published by Routledge in 2020 – against a worldwide right-wing drift towards "Our country first!" slogans. These destabilise democracies today as much as they did in the 1930s.

It was in 1963, shortly after the publication of Carson's book on ecology, that psychiatrist Frantz Fanon's "Les damnés de la terre" – to which Jean-Paul Sartre contributed a preface – was released in English [Fanon 1963]. That book on colonialism was a wake-up call that prompted movements in the metropoles to show solidarity with the anti-colonial liberation movements in the southern hemisphere. These solidarity movements have diversified into a multitude of initiatives.

The conclusion is that tipping points are imminent as long as justice is denied to people within societies as well as to peoples in the world. Exceeding those tipping points can entail unnecessary, bloody social upheavals resulting in wars, in devastation of the natural conditions of human civilisation and even in the lunification[f] of Earth.

---

[f] I already used that term in "Emergent Information". It was coined by the Austrian philosopher Leo Gabriel († 1987) when he was president of the Vienna

In view of all these developments, the three areas of global challenges suggest acceptance of the term Anthropocene to signify that era that was ushered in with the end of World War II and the deployment of nuclear weapons.[g] That very period of time revealed the dark side of human terraforming and its accompanying existential risks. Importantly, the dark side of the Anthropocene need not outbalance the bright side. To fulfil its potential and utilise the full range of options provided by the Great Bifurcation, civilisation needs to incorporate the anthropocenic situation and strengthen the positive features associated with shaping Earth. Such an anthropocenic civilisation can be qualified as a Global Sustainable Information Society (GSIS)[h]. GSIS is the goalpost to which the upper branch of the Great Bifurcation of trajectories is headed, thereby continuing the mega-evolution of sociogenesis.

3.1.2.2 *Global Sustainable Information Society (GSIS)*

A GSIS is defined as a society that demands the following properties.

(1) **Globality.** Regarding the spatio-temporal dimension, globality would be, for the first time in history, an integration at the level of all humanity.

Universitätszentrum für Friedensforschung, which helped facilitate the East-West dialogue during the Cold War.

[g] Verena Winiwarter, human ecologist and environmental historian at the University of Natural Resources and Life Sciences, Vienna, promotes the radioactive isotope ^{14}C as an "index fossil" for the Anthropocene. That isotope was created in summer 1945 by testing and dropping atom bombs. Until 1962, atmospheric nuclear tests pumped additional ^{14}C into the atmosphere, and from there it was conveyed to plants and ended up in the bodies of humans born during that time. It will be detectable in the dental enamel of human remains for thousands of years [Winiwarter 2021, 67].

[h] In 2004, I cast the normative vision of the "Global Sustainable Information Society", albeit initially under a somewhat different term in the framework of an invited lecture titled "The Great Bifurcation: A Sustainable Global Information Society or Extinction" at the University College Cork, Ireland. Later, I elaborated the critical social system theoretical details.

(2) **Sustainability.** Regarding the level of organisation, sustainability would be a reorganisation between the interdependent social systems in which humanity is currently scattered. It would also involve reorganisation within the social systems such that sociogenic dysfunctions of the cultural, political, economic, ecological and technological subsystems can be kept below a threshold that would endanger the continuation of sociogenesis. That task requires fully integrating all of humanity.
(3) **Informationality.** Regarding the state of collective intelligence, informationality would be a capacity to create information that is required for the reorganisation [Hofkirchner 2017a, 15].

Redefining society in this vein would convey a new meaning to the well-known terms globality and sustainability, and coin the neologism of informationality. These three properties are discusssed in more detail below.

Globality: an emergent spatio-temporal dimension. The drive of social systems towards globality, which is globalisation, goes beyond a mere social event created by powerful economic players and fought for by civil society movements. It also includes a seminal process grounded in a tendency typical for the evolution of complex systems: "when independent systems have become interdependent, evolution on the same level can become punctuated by the transition to a meta-system that forms a hierarchy; a supra-system can emerge that nests the interdependent systems as co-systems" [Hofkirchner 2017b, 287]. This coalesces to continue evolution (Figures 3.9.a to c).

In its solitary phase, a self-organising system n is endowed with agency and exhibits a micro- and a macro-level, which makes it capable of executing contingent activities (the upward arrow symbolises emergence). It can be regarded as a proto-element, if further phases follow.

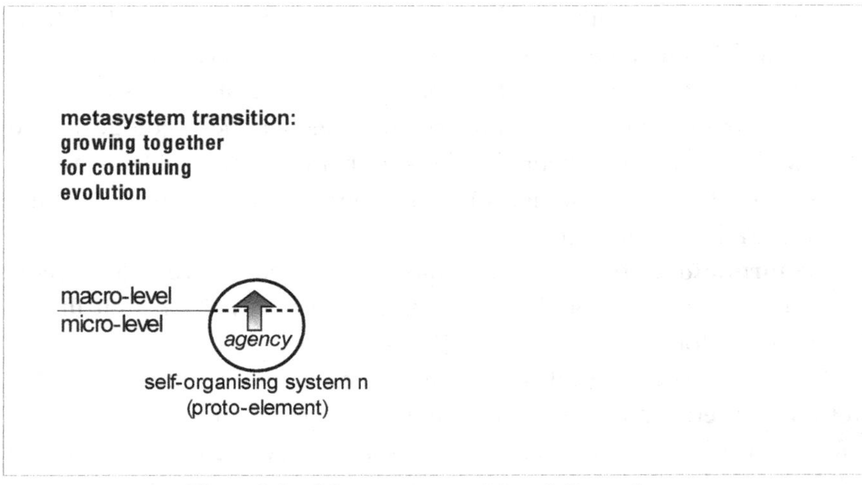

Figure 3.9.a. Metasystem transition. Solitary phase.

In the interdependency phase, self-organising systems n and n+1 interact (the up- and downward arrows symbolise the capabilities to self-regulate).

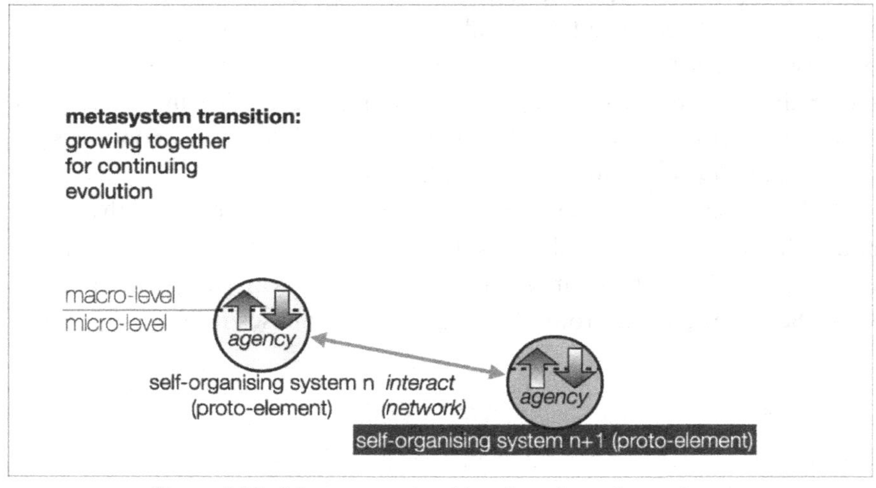

Figure 3.9.b. Metasystem transition. Interdependency phase.

In the integrative phase, the systems' interaction turns into co-action, by which they trigger the emergence of a metasystem through the emergence of relations of organisation for the metasystem. They are at the tipping point from proto-elements to elements of the metasystem.

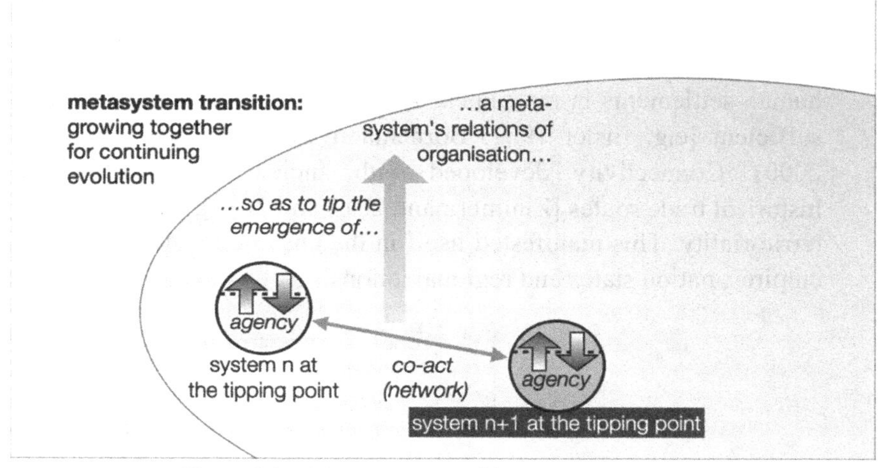

Figure 3.9.c. Metasystem transition. Integrative phase.

This tendency is recognisable during globalisation.

> Formerly rather independent social systems in different parts of the globe have been becoming more and more dependent on each other. There has been going on a penetration of social systems in range and depth, there has been less and less left over, there is almost nothing "outside" any more. This is an objective rise of interdependence between social systems. And this paves the way for the development of forms of common governance for all those systems. Such a metasystem transition is possible. Moreover, it is imperative, for what can be conceived as constitutive partitions of humanity on earth cannot survive and thrive unless all and each of them become, in fact, integrative parts of a society of societies. [Hofkirchner 2017a, 15-16]

The evolution of globality originates from two preceding stages [Hofkirchner 2017a, 16] (Figure 3.10).

(1) **Nomadism.** A first stage, which resembles the solitary phase of early human groups, is nomadism of foraging bands. At this stage, hunters and gatherers acted like animal predators. When prey was overhunted

or wild plants could not satisfy needs, the nomad migrated to other places for foraging. The encounter with other nomads was not a determining factor. Encounters may have proceeded peacefully and if rivalry occurred, there was place enough for avoidance.

(2) **Territoriality.** A second stage then developed when interaction and networking of nascent societies led to an interdependency phase. In particular, after the neolithic revolution, the first agglomerations of human settlements in matrilineal societies presumably became self-sufficient [e.g. Eisler 1987; Bornemann 1975; but see also Eller 2000]. Connectivity developed with ancient city-states along historical trade routes [Zimmermann 2014 and 2015], culminating in territoriality. This manifested itself in the ensuing Roman and other empires, nation states and regional unions.

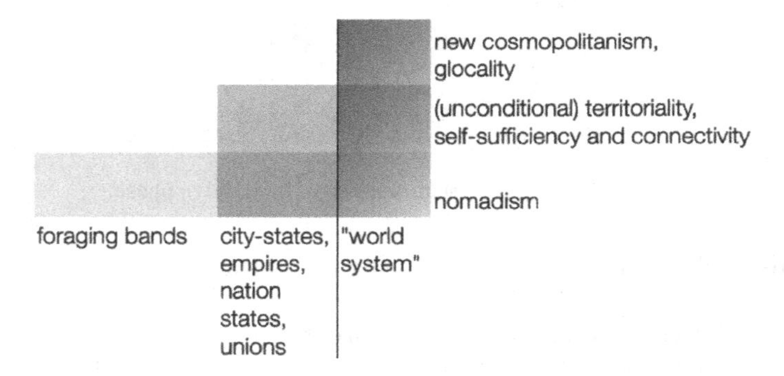

Figure 3.10. Evolution of globality. The spatio-temporal dimension of GSIS.

(3) **New cosmopolitanism?** Given these developments as preconditions, a third step seems possible that negates the exclusive feature of territoriality and resumes nomadicity [Lévy 1997], self-sufficiency and connectivity under the new circumstances. That new feature is a new cosmopolitanism, it is globality that – in the sense of R. Robertson's definition of the term glocalisation [1992] – embraces the bottom-up and top-down dynamics in a world system that – not in the sense in which Immanuel Wallerstein [1988] introduced the term – is a social system of nested social systems that run from the local level up to the global level. In that vein, nation states need not to be dissolved when it comes to the world system. They need

only being reworked. Democracy needs to be strengthened from below as far as the global level is not hampered. [Hofkirchner 2017a, 16] [i]

Thus, globalisation is a process that needs to be reclaimed as alter-globalisation in the sense of the motto “another world is possible”. Territoriality must strip off its unconditional quality in order to become fit for the glocal. Such an evolution in space and time is within reach.

Sustainability: an emergent level of organisation. From an emergent systems view, the drive towards sustainability, which can be called the process of sustainabilisation, refers to how any meta-/suprasystem gains agency itself and exercises this agency for the attainment of stability: “when the meta-system transition and the re-ontologisation of the new whole system according to the new organising relations has started, the new structure enters into operation that can enable and constrain the interaction of the new co-systems for the sake of a stable development of both the supra-system and the co-systems” [Hofkirchner 2017b, 288]. This is shaping the whole for unity-through-diversity (Figure 3.11).

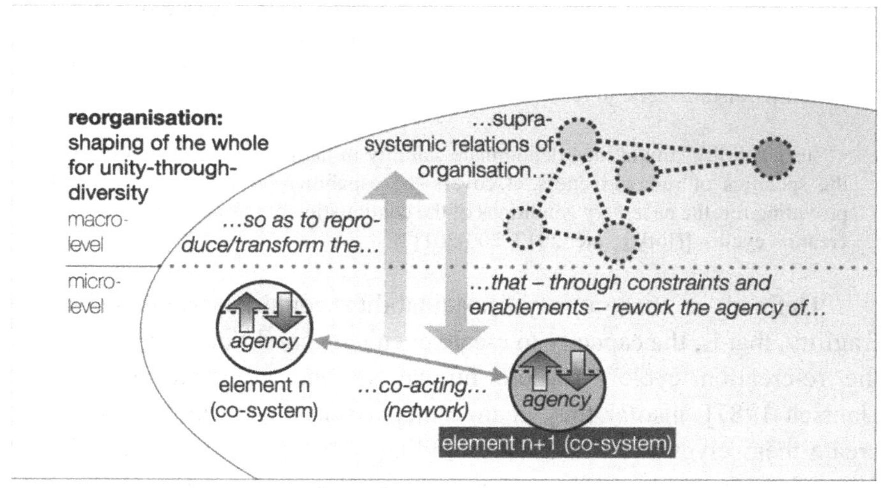

Figure 3.11. Reorganisation.

[i] Wallerstein determined the beginning of the “world system” some 500 years ago. In my view, I would reserve the term “world system” for a system that gathers the existing societal systems under the umbrella of a new transnational governance.

After having triggered the emergence of the metasystemic relations of organisation by co-acting proto-elements, the latter become fully-fledged elements. The organisational relations of the fully-fledged suprasystem now exert a dominance over the elements in that they support the elements' co-action for the further development, i.e. reproduction or transformation of the existing organisational relations.

Stability marks the steady state in self-organisation. Unity through diversity means: "there is not only unity (a suprasystem) in diversity (the co-systems) but also diversity in the unity and the unity cannot exist without diversity, unity is in any moment of its existence made up and renewed by diversity" [Hofkirchner 2017a, 17]. Unity-through-diversity is the generic stability feature of self-organising systems. It comes in several varieties:

> • "robustness" could signify stability in pure material systems; there is a threshold beneath which disturbances can be tolerated, while disturbances of greater impact can destroy the order of the system;
>
> • "resilience" – a term introduced in ecology denoting ecosystem succession cycles [Holling 1973] – can be generalised so as to designate stability in pure living systems that are able to react more flexibly than pure material systems, to bounce back after disturbances, to recover, to repair themselves;
>
> • "sustainability" might then denominate stability in human systems; according to the specifics of re-creativeness, it covers the capability of taking care of, and providing for, the necessary conditions of the continuation of the social systems' re-creation cycles. [Hofkirchner 2012, 200-201]

"Defined in such a way, sustainability comprises so-called anti-fragility, that is, the capacity to create even new conditions that better suit the re-creation cycle", "where human systems re-create themselves [Jantsch 1987], insofar they change their environment and by doing so create themselves anew in a never ending process" [Hofkirchner 2017a, 17]. Of course, sustainability refers to the ecological context.

> But there are more factors to be taken into account, if a social system is to be held within a stable bandwidth of futurable development. And awareness has developed that it is the presence of a certain kind of social relations or the absence of another kind of social relations that causes dysfunctions in the ecological realm as it does in the social and the technological realms: a society whose social relations are not

> inclusive, is not sustainable; it is not sustainable in the social respect (it cannot provide justice, equality, freedom, solidarity to a sufficient degree) and, for that reason, it is not sustainable in the eco-social respect (it cannot provide survivability to a sufficient degree) and not in the techno-social respect (it cannot provide thrivability of its productive forces to a sufficient degree) but is prone to suffering from severe dysfunctions. Since those dysfunctions are man-made it is possible, if not to eradicate them, at least to ease them to a degree that allows for keeping the social systems on a sustainable path of development. [Hofkirchner 2017a, 17-18]

As to the evolution of sustainability, again two subsequent phases with different levels of organisation can be reconstructed from history. The second one is conceived as a negation of the first one, and a third one can be postulated as the possible negation of the negation (Figure 3.12).

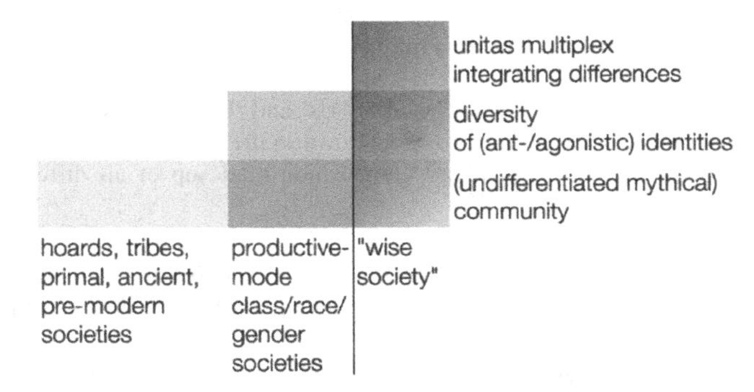

Figure 3.12. Evolution of sustainability. The level of organisation of GSIS.

(1) **Community.** The first stage is characterised by undifferentiated mythical communities as totalities in hoards, in tribes and in primal, ancient and pre-modern societies. The dialectic between the structure and the agency was mystified: a holistic We was projected onto every individual via Mead's "significant others" [Mead 1934].
"Ends are given to the community by the mythology and so are the means. Social self-organisation of those communities was always so. There is no significance attached to a differentiation between means and ends. Neither ends nor means can be questioned" [Hofkirchner 2014c, 137].

(2) **Diverse identities.** The second stage, currently still dominant, counts several types of productive-mode class societies. By producing a

surplus value, heteronomy and class/race/gender divides became possible and have become reality. It seems they have dominated the agency-structure dialectic ever since. The result is a diversity of antagonistic and agonistic identities, in which freed individuals compete against each other or form tactical coalitions.
"The final end is given: everything has to serve the selfish interests of the 'I'. However, more and more flexibility has been granted to the means" [Hofkirchner 2014c, 138].

(3) **Unitas multiplex?** The second stage can now be overcome and the means-end relationship can develop further:

> a possible third step would be the feature of a "wise society" as put forward by the High-Level Expert Group of the European Commission in 1997, namely an unitas multiplex [Morin, 1992, 143-144], a universal without totality [Lévy, 2001], integrating the differences in synergetic "Me"-"Us"-"Thee" triads – "Me"s and "Thee"s are "I"s and "You"s mediated by all of "Us" and "Us" is the "We" mediated by "Me" and "Thee", from the world system down to the most local system – that foster complementariness, a subsidiary composition made up of all differences. [Hofkirchner 2017a, 19]

> No means, no ends are given unless agreed upon in common. Not only are the means variable, but also the ends are not constant any more [Hofkirchner 2014c, 139].

The new unity allows diversity to reveal its synergistic features by stripping off its antagonistic and even agonistic ones, while the community level has become differentiated and lost its mythic. The evolution of such a level of organisation is clearly attainable.

Informationality: an emergent state of collective intelligence. A trend towards informationality – "informationalisation" [Hofkirchner 2013a, 251] is recognisable and inherent in any evolutionary system. The ability of a system to organise itself is co-extensive with the ability to generate information. "Information is generated if self-organising systems relate to some external perturbation by the spontaneous build-up of order they execute when exposed to this perturbation" [Hofkirchner 2013a, 172].

> Information is that very process of relating or the result of that process which is the order itself. Intelligence is the capability of self-organising systems to generate that information that contributes in the best way to solving problems that occur to the

> systems when maintaining themselves or improving their performance. Collective intelligence is emergent from the single intelligences of the co-systems on the level of the suprasystem. Collective intelligence can do better than any single intelligence. In times of crises, systems are prompted to organise themselves onto a higher level to overcome the crises. The better their collective intelligence, that is, the better their problem solving capacity and the better their capability to generate information, the better their handling of the crisis and the order they can reach. [Hofkirchner 2017a, 19]

This is a reformulation of W. Ross Ashby's Law of Requisite Variety [1956].

> In cybernetics, W. Ross Ashby introduced the law of requisite variety [...]. According to it, a system is said to be able to steer another system, if the variety it disposes of corresponds, if not surpasses, the variety of the system to be steered. That is, its options to (re)act shall correspond to the options the system has. If we connect variety to complexity, we can reformulate that law as follows: the steering side needs to be at least as complex as the challenge by which it is confronted from the side to be steered. If we connect complexity to information, we can arrive at the following conclusion: the steering side can increase its own complexity through generating information. Cases that apply here can include not only systems in the outer environment that are attempted to be steered but also the system itself (its inner environment). Thus, we can introduce a law of requisite information. Requisite information is that appropriate information a system has about the complexity of the exterior and interior environment. Requisite information safeguards the functioning of the system. Informationalisation is then, in contradistinction to mere informatisation, the transformational tendency towards informationality, enacted by informed actors and social systems whose capacities allow for the creation of requisite information, not least because of a suitable informatised infrastructure. [Hofkirchner 2020a, 3]

The law of requisite information can even be extended to the law of ubiquitous generability of requisite information because all complex systems have the potential to generate information. Generating requisite information is all about catching up with complexity gaps (Figure 3.13).

Complex systems are subject to – and the subject of – collective intelligence. If a complex system turns out to be objectively under-complex, it needs to undergo an information revolution, as James R. Beniger postulated [1986]. As a suprasystem, it thereby makes use of its elements. The elements are able to improve the system's intelligence by generating cognitive, communicative and co-operative information that contributes to a proper performance of the system's agency. They can cognise the organisational relations (small dotted arrows), they can communicate among them (dotted arrows) and they can co-operate with each other via the organisational relations (bold arrows). All these activities harbour the possibility to generate requisite information. Of course, they can fail to realise the possibilities in time or to the necessary extent. When applied to the reorganisation trends and metasystem transitions present in any evolutionary system treated above, that statement can be modified: a suprasystem has better chances to succeed in dealing with complex challenges than those isolated systems or systems that interact without the perspective of a metasystem transition.

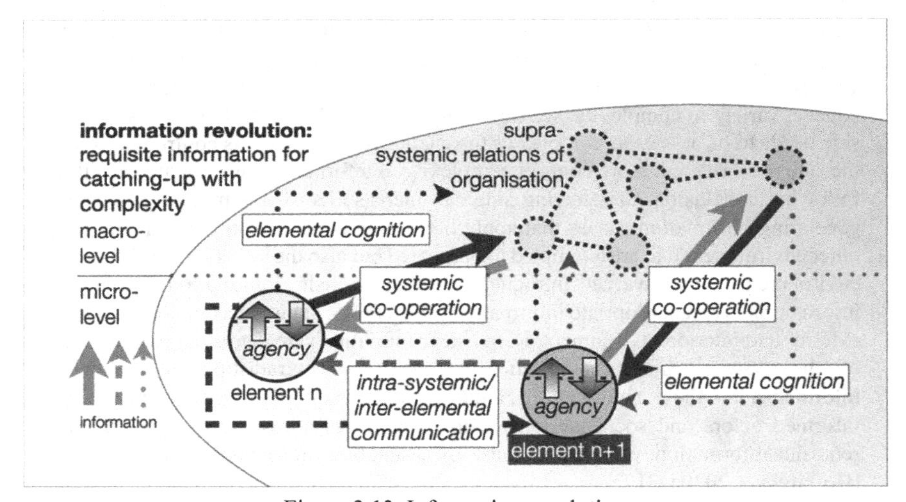

Figure 3.13. Information revolution.

This also holds for the evolution of social informationality. The law of generability of requisite information is exceedingly accurate in the case of *Homo creator*.

Applied to social systems, that law is the system theoretical expression of the law Karl Marx introduced into social science when postulating a dialectic between productive forces and relations of production and the substitution of relations of production and the whole societal superstructure when they do not meet any longer the requirements of the productive forces. Today, global challenges are complex and need complex solutions. Complex solutions would be possible, if humanity developed that information that is necessary to shift societal development onto a sustainable path. [Hofkirchner 2017a, 20]

A three-step logic can start with the following sketch (Figure 3.14).

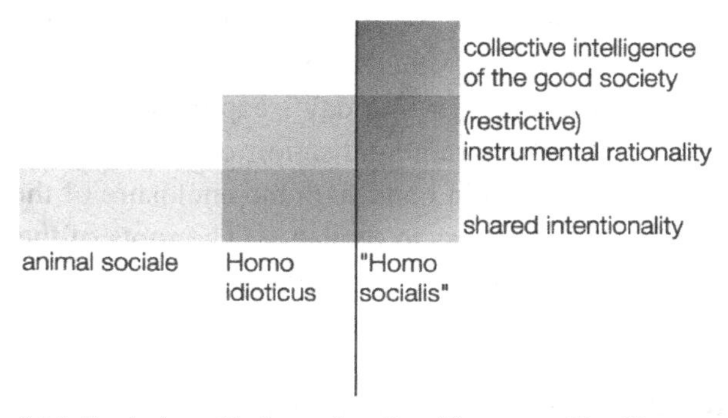

Figure 3.14. Evolution of informationality. The state of intelligence of GSIS.

(1) **Shared intentionality.** When the transition from the last common ancestors of great apes and humans started, social factors were inserted into the biotic evolution – an *animal sociale* was formed with shared intentionality. As illustrated by Tomasello's stag hunt example described above in section 2.2.1, the co-operative turn in evolution triggered the coevolution of joint goals, joint actions and joint attention.

(2) **Instrumental rationality.** At some point, dyads became overlaid with triads. Greek antiquity's advent of cunning is demonstrated in the interpretation of the Odyssey [Holling and Kempin 1989, 17-31] following Frankfurt School's Dialectic of Enlightenment [Horkheimer and Adorno 2016]. That interpretation highlights, however, that triads were hijacked by the instrumental rationality of *Homo occidentalis* [Bammé 2011] and *Homo economicus*. This yielded a new stage of intelligence – *Homo idioticus*[j]. The Auckland media lecturer Neil Curtis writes that "the creation of the private through the enclosure of public or commonly held resources has historically been the primary means by which property has been

[j] My term.

secured for private use". As etymology shows, in antiquity the Greek "idios" meant "the personal realm, that which is private, and one's own", and it also bears the stamp of "that which is enclosed" [Curtis 2013, 12]. The term "idiotes", then, denoted a person concerned with his/her personal realm only, with his/her own, and not with, say, the *res publica* and the fate of other human beings. "Idiotism" is a term that Curtis uses as a signifier of today's capitalism and neoliberalism – as institutionalised instrumentalisation of humans for one's own interests. This goes hand in hand with the enclosure of the commons and the denial of free access to the latter. The roots of the term were laid by Greek patriarchs, who enjoyed private property and conversed, as autonomous subjects, with each other as intersubjective identities [Holling and Kempin 1989, 30-31].

(3) **Collective intelligence of the good society?** What about the third step? It can open the perspective of a new collective intelligence, implanted in the good society:

> Third, when collective intentionality would be set free from the current restrictions that instrumentalise co-operation for competition against other "Them"-groups, a true *"homo socialis"* [Gintis and Helbing, 2015] could enter the stage. New cosmopolitans could enjoy a universalised, extended capability to act through a consciousness and conscience that takes care of the global commons. [Hofkirchner 2017a, 21]

Instrumental rationality can overcome its restrictive state and become a useful part of the collective intelligence of the good society. Such a state of societal intelligence is feasible.

Together with a new organisational level and a new spatio-temporal dimension, all three features of a GSIS can be implemented in a third step of social evolution – as a Third that builds upon the two steps of hominisation that history has passed through and continues with a higher-order humanisation.

A new world-culture is within reach.

3.1.3 *Reframing the social in the age of the Great Bifurcation: the emergence of anthropo-relationality*

Relationality is changing too in the Anthropocene. A gap is evident between the behaviour that appears glocally and the global relations that are essential for understanding and for drawing practical conclusions (but are hidden).

The individualistic view asserts that global relations result from glocal behaviour, which is a reductionist mistake because global relations exhibit a new quality that the glocal behaviour does not have. If the reductionist view were true, social relations would be indistinguishable from social behaviour. The projective turnaround cannot be true either. Behaviour is not a result of the relations, although relations represent limits of behaviour. Global relationality provides social conditions for the behaviour in different locations, and the behaviour more or less complies with the conditions.

Moreover, both aspects cannot be independent of one another. This fallacy would disconnect the superficial from the integral and have no explanatory value.

Table 3.3. The consideration of the apparent and the essential in bifurcationist epistemology. The (precautionary) glocal behavioural and the global/anthropo-relational.

		apparent	**essential**
con-flation	**reduction: individualistic fallacy**	**the glocal behavioural:** sufficient condition for the global relational	**the global relational:** resultant of the glocal behavioural
	projection: sociologistic fallacy	**the glocal behavioural:** resultant of the global relational	**the global relational:** sufficient condition for the glocal behavioural
disconnection: anarchist fallacy		**glocal behavioural**	**global relational**
		incommensurable knowledge	
combination: bifurcationist epistemology at the Great Bifurcation		**the precautionary behavioural:** necessary condition for an anthropo-relational	**the anthropo-relational:** an emergent from the precautionary behavioural

The gap is bridged by bifurcationist epistemology. The necessary condition for a real global relationality that can emerge from glocal behaviour around the world is precaution [Jonas 1984] when dealing with the anthropogenic problems of the Anthropocene. And such real precaution can to a certain extent trigger what is meant by the term anthropo-relationality [Barthlott et al. 2009; Deutsches Referenzzentrum für Ethik in den Biowissenschaften].

The emergence of existential risks signifies an evolutionary crisis of complex social systems. Those risks render dysfunctional the prevailing logics that up until now have underlain the social relations as objective logics in reality. The antagonisms and agonisms of restrictive-intelligent *Homo idioticus* in territorial states of class/race/gender societies (due to enclosures of cultural, political, economic, ecological and technological commons) all appeared in the second stage of sociogenesis. Although they became exacerbated over time, they did not hinder progress from being made. Nonetheless, given the new requirements for sharing the *commune bonum* under the conditions of globality, sustainability and informationality in the Anthropocene, they constrict further progress. They prove to be anachronistic and constrictive in all the realms that social relations cover according to the Critical Social Systems Theory (Table 3.4):

Table 3.4. Categories of social relations and their logics. Second step of sociogenesis and a possible third step.

		sociogenesis	
		logics of the second step	**logics for a third step**
social relations	**among humans**	self-centricity: exceptionalism	pan-humanism: united humanity
	with nature	anthropocentricity: colonisation	anthropo-relational humanism: alliance with nature
	to techno-logy	power-centricity: super empowerment	digital humanism: ethically aligned technology

(1) **Human-human relations.** These refer to social relations that tie humans with and via society (social relations in the proper sense).

(2) **Human-nature relations.** Those natural social relations connect humans and nature.
(3) **Human-technology relations.** Those technological social relations link humans to technology.

3.1.3.1 *The logics of human-human relations*

Social relations proper make up the generic category. They determine the other categories.

Self-centricity. The social relations among humans continue to be determined by a logic of self-centricity. Social self-centricity means the centredness of a social self of whatever kind on itself. Social selves are social systems, which in most cases are selves within another social system. It signifies the denial of belonging of a social partition to a greater social whole, whatever that partition may be – a nation, an ethnic group, a private enterprise or an individual, etc. That denial equals exceptionalism in that the particular is esteemed vis-à-vis other particulars so as to hijack the universal. Importantly, such a logic has incrementally deprived the growing masses of the world population (as well as partitions of populations) of access to the societal common good. It is a logic of domination, exploitation and oppression in the anthroposphere and has yielded a gap between rich and poor, hunger, diseases, and much more.

Pan-humanism? It is crucial that self-centric exceptionalism of the relations among humans give way to the logic of pan-humanism [Morin 2021]. Pan-humanism can be defined as a logic of the single integrated, united humanity, of living together for the good of all. That logic is inclusive. It does not exclude any part of common humankind by antagonisms (zero-sum plays that benefit some at the cost of others). Rather, it includes all of them in synergisms, in compositions of parts that let them share any benefit achieved by any part. Being an objective community of destiny denies humanity any rationality for competition that does not serve the co-operation for the common good.

Pan-humanism can be understood as a kind of anthropo-relationality. Every subject is a relational subject (see section 2.2.3). Either is the subject humanity itself constituted by a plethora of human subjects, or it is one of the human subjects that relate him-/herself to humanity.

Pan-humanism would also be the remedy for the current pandemic situation because a zero-COVID strategy is only implementable if all countries of the world are put in the position to fight the virus by sufficient vaccination. No country can reach a zero-COVID state if the remaining countries are not committed to the same policy, just like no single person can be protected from infection unless a sufficient number of persons is vaccinated, without free riders who frustrate solidarity.

Pan-humanism includes the concern that the next generations be provided with at least the same developmental potential as the current generation. It therefore implies principles for the social relations with nature and technology in the sense of the German philosopher Hans Jonas [1984].

3.1.3.2 *The logics of human-nature relations*

Social relations with nature are the second category. They depend on the first category.

Anthropocentricity. Self-centricity of humans has morphed into anthropocentricity when nature is involved. Anthropocentricity does not include natural systems. As long as the social relations among humans remain self-centric, the social relations with nature are determined by a logic of colonisation [Fischer-Kowalski and Haberl 1993]. As long as social systems neglect other co-existing social systems, they also fail to take care of co-existing systems of natural origin. Such an anthropocentric logic has incrementally undermined the ecological foundations of human life on Earth. It is a logic of depletion and contamination of the biosphere and the geosphere, yielding a decrease in biodiversity, an increase of the heating of the planet, and much more.

Alliance with nature? Anthropocentric colonisation of nature must give way to a logic of alliance with nature – a phrase coined by German philosopher Ernst Bloch [Zimmermann 2012]. Alliance with nature is anthropo-relational. This means that humans relate to nature by giving up their anthropocentric perspective of colonisation, though not giving up their specific position of an *animal sociale* or *zoon politikon* that makes them distinct from agency in pure nature. The social relations to nature need to shed their anthropocentricity but not humanism as a whole.

Humanism must become an anthropo-relational humanism. In a systemic perspective, humans as self-organising systems need to concede self-organising capacities to non-human natural self-organising systems too when integrating them with their pan-human social systems. While humans need to concede those capacities to the agency of non-human systems, this must proceed in a staged way. If humans wish to remain in compliance with their own place in evolution, it must reflect the objective place of the non-human systems in physical and biological evolution. Thus, humans are prompted to relativise their own positions. In considering human anthropo-relationality, social relations with nature need to do justice to the place of natural systems. If the objective function of a pan-human system is the global common good, then the objective function of an alliance with nature is also a global common good. It shares the productivity of society with the productivity of nature – a relation that is in itself a commons.

Again, the COVID-19 pandemic is an example for avoiding risks for humanity by not repressing or intruding into wildlife nature when applying the alliance-with-nature perspective.

Alliance with nature implies principles that are valid for the social relations to technology.

3.1.3.3 *The logics of human-technology relations*

Social relations to technology are the third category. They depend on the second category.

Power-centricity. Anthropocentricity has taken the form of power-centricity when relating to technology. As long as the social relations with nature remain colonising, the social relations with technology are determined by a logic of super empowerment. This is the illusion of having super powers. As long as social systems fail to mind co-existing systems of natural origin, they tend to produce and use tools that are not designed to take care of those systems. Such a logic has incrementally hypostatised the effectivity of technology beyond any rational measure. A logic of omnipotence is ascribed to the technosphere. The result? The deployment of intended nuclear first-strike capabilities, the use of chemical weapons, waging information wars, the development of autonomous artificial intelligence, surveillance, trans- and post human developments, and much more.

Digital Humanism? Power-centric super empowerment has to give way to a logic of Digital Humanism [Nida-Rümelin and Weidenfeld 2018, Vienna Manifesto on Digital Humanism]. This would be practiced by human-centred technology. Digital Humanism is the logic of civilisational self-limitation, as Austrian-born writer Ivan Illich coined it [1973] – a limitation of the technological tools to their role of serving alliance-with-nature and pan-human purposes only. The observance of the precautionary principle – the "Prevalence of the Bad over the Good Prognosis" [Jonas 1984, 31] – is a *sine qua non.* Digital Humanism means a humane digitalisation. Digitalisation can provide solutions for augmenting those purposes because any information technology helps smoothen frictions in the functioning of any other technology. Regardless, digitalisation must be ethically designed and the tools cultivated, which runs under the label of human-centred technology.

Unfortunately, during the COVID-19 pandemic the precautionary principle tended to be belittled. Though some states pursued zero-COVID strategies, in many other states the politicians, economic interests and misinformed people hindered the acceptance of recommendations of scientists who supported the decision-making with computer simulations (and whose forecasts mostly came true). Far from being omnipotent in reality, potentials for the good were renounced. Many states had to experience the limits of their health services, which were unprepared for a pandemic despite anticipating warnings.

Digital Humanism is anthropo-relational too. It relates the human and the digital by relinking the tools back to the good society and its tasks.

In all human-human, human-nature and human-technology relationships, the days of prevailing self-centric, anthropo-centric and power-centric logics are waning. Those logics fail to obey the Logics of the Third. At best, they can be qualified as intermediate stages, that is, between a logic of the First – centred on the individual actor – and a logic of the Second – centred on interacting actors. A new era marked by a prevalence of pan-humanistic, anthropo-relational humanistic and digital humanistic logics is arising. It centres on the generic feature of social relations as a proper Third, which will pave the way for a next step in social evolution.

3.1.4 *The Principle of Planetarism*

Critical emergentist bifurcationism is the theoretical approach to the investigation of sociogenesis in the age of the Great Bifurcation. It fleshes out Critical Social System Theory (CSST) to make it fit for today's ongoing changes and include the Principle of Planetarism.

The appearance of global challenges calls for the analysis and synthesis of how to overhaul the design, model and frame of that which is social. That effort substantiates the claim for a planetary perspective. Such a planetarism is required to understand that, on the one hand, social evolution needs to adapt to the planetary conditions to master the evolutionary crisis and that, on the other, it is also able to implement the adaptation. The ambivalence of the social – of progress or regress, of spiralling up or spiralling down – has been aggravated with the advent of the Great Bifurcation. That ambivalence, however, by no means eliminates the objective potentials for a better future. Those potentials can be envisioned at the praxiological, ontological and epistemological levels. Contested commons are objectively set on a planetary scale for shared values in the cultural system and not for exclusive hegemony; for democratic decision-making in the political system and not for repressive strength; for common wealth in the economic system and not for fortune; for harmony in the ecological system and not for colonisation of nature; for productive forces and not for destructive ones. The features of a future world society are objectively set as a Global Sustainable Information Society – for a new cosmopolitanism in the spatio-temporal dimension, for *unitas multiplex* integrating the diverse on the new organisational level, and for a collective intelligence that a good society deserves as its new state of intelligence. The structure is objectively set to pan-human social relations, not to self-centric ones among humans; with regard to natural social relations, it is set to an alliance with nature, not to an anthropocentric colonisation of nature; and it is set to Digital Humanism, not to power-centric social relations to technology.

Critical bifurcationism highlights those potentials and claims their objective existence in reality. Those potentials pose an objective condition for a Great Transformation:

> The becoming of humans and humanity is not yet finished. No trans- or post-humanism that focus on the individual are needed. To cope with the global challenges that put our civilised existence at stake global citizens are needed. If global citizens succeed in coping with the challenges (and transform our societies into a single Global Sustainable Information Society as meta-/suprasystem), humanity would accomplish the third step to anthroposociogenesis. [Hofkirchner 2020b, 3]

This has consequences for sciences to become humanistic and for humanism itself.

Humanism revisited, update III. A humanist view of social evolution is too abstract if it neglects the centrality of commons. It is also too abstract if it neglects that the enclosure of the commons has become the ultimate cause of an existential risk for future humanity. The enclosure of the commons has reached an extent that forfeits further progress in social evolution and threatens a regression down to the loss of the potentiality that humans have been endowed with from the outset. This makes it the first and foremost agenda of critical humanism to reflect upon the fate of the human race on Earth, to provide guidance in times of crisis, and to fight anti-humanistic ideologies. Today, by diverting from the focus on the commons those ideologies pose a greater threat than ever.

Humanism is planetary.

It can be defined:

Bifurcationism. *Bifurcationism is that critical social elaboration of* weltanschauung, *that critical conception of the social world and that critical social-scientific way of creating knowledge that devises the criticist Principle of Planetarism by applying the criticist Principle of Commonism to the circumstances of the age of the Great Bifurcation.*

The **Principle of Planetarism** states: there are, regardless of whether or not thematised, associated with the entry of social systems into the age of the Great Bifurcation,

(1) the emergence of the common-destiny humanity as a form of global humanity from the glocal differences that respond to being affected;

(2) the emergence of anthropocenic civilisation as a form of a global civilisation from a glocal terra-forming that also responds to the conditions of the Anthropocene;

(3) the emergence of anthropo-relationality of the integral structure of the global relations from a superficial glocal behaviour that responds precautionarily;

such that the common-destiny humane, the anthropocenic civilising, and the anthropo-relational – thereby specifying the common goodness, the social, and the structural to-be-explored of the sociogenetic approach – are instantiations in a hierarchy of levels. At the same time, each is an emergent meta-level on its own. Accordingly, each is a Third that represents an essential property of social self-organised systems, actualised in ideations or materialisations by the interaction networks of the actors as plural Seconds, in which the participating actors – humans or other social systems – represent single Firsts. Any social system in the age of the Great Bifurcation is exposed to these required specifications of social self-organisation.

Humanism is instantiated as being planetary.

3.2 Rethinking Social Information in the Age of a Great Transformation: Emergentist Transformationism

This subchapter returns to the issue of a Critical Information Society Theory (CIST), albeit specified for the new age of a possible Great Transformation (which is simply another name for the age of the real Great Bifurcation). Critical emergentist noogenetics, with its Principle of Eudaimonism, is about the subjective conditions for progress anywhere, anytime. Critical emergentist bifurcationism, with its Principle of Planetarism, deals with the objective conditions of a Great Transformation in the age of the Great Bifurcation. This subchapter, in turn, focusses on the subjective conditions for a Great Transformation. It applies, and is based upon, both noogenetics and bifurcationism in order to yield critical emergentist tranformationism with the Principle of Convivialism.

It does so by resorting to convivialist theory. The occupation with that theory has become an academic and an international movement. It started in the last decade, when about forty mainly French intellectuals – among them Alain Caillé, Serge Latouche, Edgar Morin or Chantal Mouffe – opened the discussion on a political manifesto to redesign social relations. The first manifesto from 2013 was followed in 2020 by a second, updated one, this time edited by The Convivialist International. The first manifesto was subtitled "A Declaration of Interdependence", the subtitle of the second is "Towards a Post-Neoliberal World". Convivialism refers to conviviality, a term introduced into the academic discourse by Ivan Illich, who published a book titled "Tools for Conviviality" [1973]. "Convivial" has Latin origins and means the quality of living together in the manner of dining together (*convivor*) of hosts (*convivatores*) and guests (*convivae*) at joint feasts (*convivia*). In the second convivialist manifesto, the authors write:

> Convivialism is the name given to everything that in doctrines and wisdom, existing or past, secular or religious, contributes to the search for principles that allow human beings to compete without massacring each other in order to cooperate better: to advance us as human beings in full awareness of the finiteness of natural resources and in a shared concern for the care of the world. A philosophy of the art of living together, it is not a new doctrine that would replace others by claiming to cancel them or radically overcome them. It is the movement of their mutual questioning based on a sense of extreme urgency in the face of multiple threats to the future of humanity. It intends to retain the most precious principles enshrined in the doctrines and wisdom that were handed down to us. [Convivialist International 2020, 7]

The convivialist movement fulfils the task of preparing the subjective condition for a Great Transformation. It helps to generate the information that is required to catch up with the complexity that social systems are confronted with. There are three objective options:

(1) a system that possesses higher complexity is able to integrate systems of lower complexity into its own self-organisation;
(2) a system of lower complexity that faces more highly complex systems is able to measure up to them or even overtake them;

(3) systems of equal complexity that exist next to each other are able to organise together a supra-system that possesses a complexity higher than the complexity of either alone, which, is ultimately beneficial to each of them.

In order to realise one of the options, social systems must take subjective action and increase information. Increase of information is an emergent process, not predictable, not predetermined. Not every information created can contribute to the increase in complexity. In times of crisis, however, evolutionary systems show search processes in every direction to quickly generate new information. This holds true for social systems as well. Human creativity is seemingly infinite. Convivialism is an ideational creation that is part of such a search process.

New information is advantageous to identify tipping points that would trigger spiralling-down developments and deepen the crisis. That process must be avoided. At the same time, new information can help identify tipping points that would trigger spiralling-up developments and potentially lead out of the crisis [Thurner 2020]. Such information must gain a foothold on a critical mass of actors. This is because revolutions in social evolution do not emerge unless a quorum is reached. The informedness of such a quorum is the tipping point for the emergence of the meta-/suprasystem transition that achieves a Great Transformation. It is the subjective condition of a Concrete Utopia after Ernst Bloch [1967] or Real Utopias after Erik Olin Wright [2010]. Critical utopian thinking resembles, in the multi-stage model that lies behind the transformation of social systems, dialectic thinking of "sublation".

> The first connotation of sublation which is to break, to cancel, to nullify, that is, to discontinue, is reflected in the stage model by the point that marks the end of a certain stage of evolution. The second connotation which is to keep, to save, to preserve, to store, that is, to continue, comes to the fore when the scheme concedes that each new layer is built upon a preceding one and that the new stage comprises not only the new layer but parts of the old one. The third connotation which is to raise, to lift, that is, to leap in quality, is depicted by the notion of the higher level that exerts downward causation onto the lower ones.
>
> [...] Being critical can be ascribed to this theoretical framework in that it is normative while doing justice to the factual at the same time. For it includes not only an account of the potential that is given with the actual but also an evaluation

> of the potential which sorts out the desired. Thus this theory embraces an ascendence from the potential given now to the actual to be established in the future as well as an ascendence from the less good now to the better then which altogether yields the Not-Yet in critical theorist Ernst Bloch's sense (see e.g. Bloch 1967). These processes aimed at the Not-Yet are at the core of the dynamic of social self-organization [...]. By the notion of the Not-Yet Bloch tried to salvage the idea of utopia – it is not any longer a nowhere deprived of the possibility to get there but a future that can be glimpsed and anticipated in what is already possible here and now. [Hofkirchner 2007, 479-480]

Subsequent to Figures 3.1.a and b, Figure 3.15 illustrates Bloch's Not-yet for the case of a Great Transformation into a Global Sustainable Information Society (GSIS). Critical utopia is grounded in the space of possibilities at the present real. In a seminal state, those possibilities need to represent relations social that are universalisable as a global commons structure. They flash up at the present real as not-yet actualised. They are anticipated for an expected future real as something that is evaluated better than the actual (which is evaluated as improvable). The techno-eco-social transformation to be actualised is expected to usher in the phase of a post-revolutionary evolution in which the interactions of the social systems will be transformed into an intra-action of a GSIS as an overarching new system. In that system, the new commons structure will transform the actors into new commoning actors that, in turn, will reproduce the structure by further transforming it.

The structure is the Third. The subjective condition for this revolution is that the Not-yet needs to be anticipated. Moreover, the imagined mutual actualisation of structure and actors needs to shine forth from what can so far only be expected (hence the grey colour of the respective arrows). The subjective condition means:

> [...] global citizens need to reflect on the establishment of a higher-order world system through transnational relations that respect the social, ecological and technological commons on a planetary scale. Such relations are that Third global citizens need to design today. [Hofkirchner 2020b, 3]

The proper handling of the Not-yet constitutes a measure in accordance with the real Logic of the Third, and the Logic of the Third is a manner of thinking systemically. Accordingly, the faculty of systemic thinking, whether conscious or unconscious, is the foundation of a Great Transformation.

> Systemic thinking needs to focus on future social relations that are not yet actualised. It needs to anticipate them ideationally on a new meta-level and it needs to anticipate the meta-/suprasystem transition of the social systems. Thus, the Third is a conjecture to be devised in order to represent a solution to real-world problems.
>
> [...] Systemic thinking does not only need to anticipate what is desirable but needs to explore that that which is desirable is also possible in the here and now. Only what is potential can be actualised. Thus, it looks in the space of possibilities now for the foreshadowing of something that might become a future Third [...]. [Hofkirchner 2020b, 3-4]

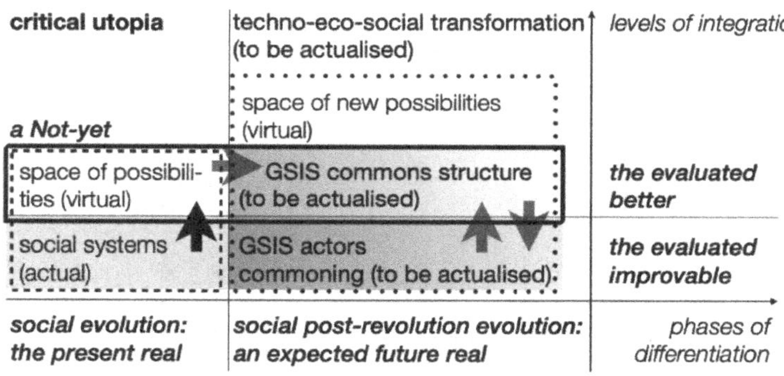

Figure 3.15. A Great Transformation for an emergent GSIS. Critical utopia.

The noogenetic constants and the bifurcationist items are treated in the next four sub-subchapters, implementing the following considerations:

> Systems thinking has been a companion of humanity from the outset. Triadic co-operation, a living together according to social relations on the system's macro-level, required the reference to a Third, a normative meta-level, that mediated the entanglement of actors, which required communication about this entanglement by coupled actors, which required the (re)cognition of this entanglement and this coupling in their minds.
>
> Today, systems thinking needs a next step of development to guide the next step of social evolution. [Hofkirchner 2020b, 4]

Convivial transformationism is that perspective, conception and way of thinking in the age of the Great Bifurcation that applies the new paradigm to the complex challenges facing global social-informational systems. The question is: What are the potentials inherent in the noogenetic constants – consensual normativity, consilient discursivity, conceptual reflexivity – that can be employed to carry out a Great Transformation? The answer elaborates another piece of transformation science: there are detectable and cultivatable potentials of a planetary ethos for global governance, of a planetary agreeableness for a global dialogue, and a planetary mindfulness for a global citizenship (Figure 3.16).

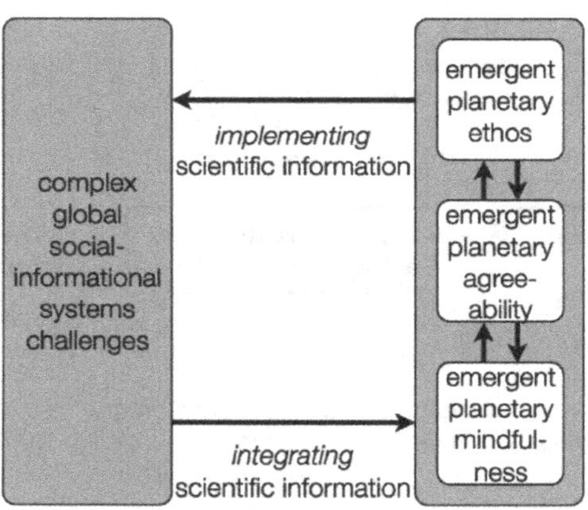

Figure 3.16. Critical emergentist-systemist-informationist transformationism. The emergent convivial content of the new paradigm.

3.2.1 *Redesigning social information in the age of a Great Transformation: the emergence of a planetary ethos for global governance*

"Is and Ought" of the social, given the Great Bifurcation, produces humanity as an objective community of common destiny, triggered by all the diversity that is objectively affected. "Is" and "Ought" of social information, then, question how the interdependency of subjects across the world can trigger a true global normative subject (Table 3.5).

No individualistic answer would accomplish the conceptual leap from the world-wide interdependent to the globally normative because it would retain the concepts on a flat plain as would sociologism (which does this on a higher level).

An anarchist answer would situate each of the two concepts in a different world.

The answer transformative praxiology gives is: planetary conscience can emerge from the Great-Bifurcational interdependence.

Table 3.5. The consideration of Is and Ought in transformationist praxiology. The (Great Bifurcation) wide interdependent and the globally/planetary conscientious normative.

		is	**ought**
con-flation	**reduction: individualistic fallacy**	**the wide interdependent:** sufficient condition for the globally normative	**the globally normative:** resultant of the wide interdependent
	projection: sociologistic fallacy	**the wide interdependent:** resultant of the globally normative	**the globally normative:** sufficient condition for the wide interdependent
disconnection: anarchist fallacy		**wide interdependent**	**globally normative**
		disparate takes	
combination: transformationist praxiology for a Great Transformation		**the Great-Bifurcation interdependent:** necessary condition for a planetary ethical normative	**the planetary ethical normative:** an emergent from the Great-Bifurcation interdependent

The first Convivialist Manifesto already incorporated the issue of interdependence in the subtitle: "A declaration of interdependence" [Convivialist Manifesto 2014]. This declaration sees the root of all threats in the following problem:

> Humankind has achieved astonishing technical and scientific feats but has remained as incapable as ever of resolving its fundamental problem, namely how to manage rivalry and violence between human beings. How to get them to cooperate – so that they can develop and each give the best of themselves – and at the same time enable them to compete with one another without resorting to mutual slaughter. How to

> halt the now limitless and potentially self-annihilating accumulation of power over humankind and nature. [Convivialist Manifesto 2014, 23-24]

The convivialist perspective is itself normative and shares the emancipatory claim [Adloff and Costa 2020]. It asks "What is the most precious thing? And how can it be defined and understood?" "A philosophy of the art of living together, it is not a new doctrine that would replace others". "It intends to retain the most precious principles enshrined in the doctrines and wisdom that were handed down to us" [Convivialist International 2020].

> There is, however, a definitive criterion instructing us as to what we can retain from each doctrine in a perspective of universalization (or pluriversalization), taking into account both the threat of possible disaster and the hope for a better future. It is to be retained for sure from each doctrine: what makes it possible to understand how to control excess and conflict so that they do not turn violent; what encourages cooperation; and what opens the way to dialogue and the confrontation of ideas within the framework of an ethics of discussion. [Convivialist International 2020]

In order "to draw the general outlines of a universalizable doctrine", the first manifesto postulated four principles that were extended, in the second one, to five plus one imperative for the "only legitimate policies, but also the only acceptable ethics" [Convivialist International 2020]. The first principle ran in the first manifesto as the principle of common humanity:

> Beyond differences in skin-colour, nationality, language, culture, religion and wealth, gender and sexual orientation, there is only one humanity, and that humanity must be respected in the person of each of its members. [Convivialist Manifesto 2014, 30]

The second manifesto adds to that point the importance of the feeling of belonging to a global human community: "A sense of solidarity will be shared among millions, tens and hundreds of millions, even billions of people, from all countries, all languages, all cultures and religions, all social conditions, driving them to participate in the same struggle for a fully humanized world" [Convivialist International].

3.2.1.1 *The imperative of a planetary ethos*

Such a global human community does not need to exist solely as an objective destiny. It can also be completed by construing a subjective social-informational backbone that reflects the requirements of a common future. Such a backbone would constitute a planetary ethos (Figure 3.17).

The figure shows three stages. Beyond the first and the second stage, a possible third stage is inserted. The first and the second stage correspond to the dyadic and triadic stages of sociogenesis after Tomasello [2014; 2016; 2019], the third one hypothesises the emergence of a so-called omniadic stage. That involves the iteration of the emergence of Tomasello's triad – a universalisation of the Third of the second phase, elevating it to the planetary level. The third stage represents a possible global phase on a possible global level.

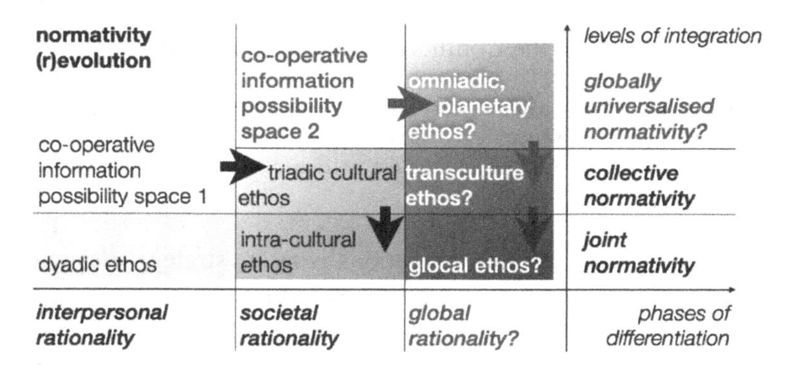

Figure 3.17. Normativity (r)evolution. Leap to a planetary ethos?

The x-axis depicts three phases of rationality: the interpersonal phase, the societal phase and a potential global one. The y-axis describes three levels of normativity: the joint level, the collective level and a potential globally universalised level. The first stage presents a plurality of dyadic ethos on the lowest level. The second stage is differentiated into a triadic ethos of a plurality of different cultures on the next-higher level, and into an intra-cultural ethos as a reworking of purely dyadic ethos. Finally, the potential third stage is once again differentiated into an omniadic, planetary ethos according to an even higher level, into a transcultural ethos as a reworking of the triadic cultural ethos according to the now intermediate level, and a glocal ethos as a reworking of the intra-cultural ethos according the lowest level. The arrows indicate the implied transformations. Arrows for the past and current stages are in black, arrows concerning the potential third stage are shaded in grey. The third stage is envisioned as emerging out of the space of co-operation possibilities of the second stage, just as the second stage emerged out of the possibility space of the first stage.

The graphic introduces the issue of rationality because ethos must go beyond being a matter of feeling only. Convivial normativity can be established on rational grounds. An example is Chinese philosopher Tingyang Zhao's endorsement of the so-called "Confucian Improvement". That refers to an inclusive Pareto improvement for everybody in the framework of the concept of "Tianxia", translated as "all-under-heaven", a theory of world politics developed about three thousand years ago [Zhao 2019; 2020]. He defines the Confucian Improvement as follows:

> [...] if Party x gets interest gains x+, then and only then will Party y get simultaneous interest gain y+ and vice versa. Hence, promoting x+ becomes a beneficial strategy for y because y must recognize and promote x+ in order to get y+ and vice versa. [Zhao 2019, 64]

Such a definition enables operationalising a strategy designed to maximise reciprocal interests and goes beyond a unilateral improvement that merely "seeks to maximize self-interests and in so doing only aggravates conflicts and confrontation" [Zhao 2019, 59]. This ensures coexistence as a necessary condition for existence, and minimising mutual hostility then takes precedence over the maximisation of self-interests – both components can develop convivial rules.

The terms "joint" and "collective" in the graphic refer to Tomasello's distinction of intentionality. The term "universalised" is used in the sense that convivialism refers to "universalization (or pluriversalization)" [Convivialist International 2020].

The term "transculture" goes back to Wolfgang Welsch's idea that cultures penetrate each other, thus rendering each other hybrid [1999].

A planetary ethos would be home to universal human values, norms and interests. Such a home of world-embracing morals and ethics provides an ideational foundation to the social relations of the world society. However, a planetary ethos would consist of more than the ethical foundations of social relations, relating the conduct among social systems and actors worldwide. It would even include the ideational foundations for the new social relations with nature and to technology, because the latter would tend to become consistent with the first.

As regards the social relations with nature, the second convivialist manifesto stated (as the first of five principles, before the principle of common humanity) the principle of common naturality:

> Humans do not live outside a nature, of which they should become "masters and possessors." Like all living beings, they are part of it and are interdependent with it. They have a responsibility to take care of it. If they do not respect it, it is their ethical and physical survival that is at risk. [Convivialist International 2020]

That principle represents a subjective formulation for the required objective logic of an anthropo-relational alliance with nature. If it states that humans are part of nature, then it must also be accepted that nature is part of humans, or better, natural systems are part of social systems. This is an evolutionary result. That which comes later and is a specification of the former is that which tips the scales by triggering the transformation. Thus, just as the biosphere is encapsulated in the noosphere, nature is nested by (world) society. Natural systems have intrinsic values and the system of humanity must accept that; it must take those values into consideration when boosting the social self-organisation with natural self-organisation.

Regarding technology, convivialist manifestos did not formulate a principle or an imperative. Although the origin of the academic use of the term goes back to Illich's book "Tools for Conviviality". In his book, Illich submitted

> the concept of a multidimensional balance of human life which can serve as a framework for evaluating man's relation to his tools. In each of several dimensions of this balance it is possible to identify a natural scale. [Illich 1973, x]
>
> Once these limits are recognized, it becomes possible to articulate the triadic relationship between, persons, tools, and a new collectivity. *Such a Society, in which modern technologies serve politically interrelated individuals rather than*

> *managers, I will call 'convivial.'* [...] *I have chosen 'convivial' as a technical term to designate a modern society of responsibly limited tools.* [Illich 1973, xii]

The German translation of the book was titled "Selbstbegrenzung" [1975], which means self-limitation. The growth of tools "beyond a certain point increases regimentation, dependence, exploitation, and impotence" [1973, 20]. "Convivial tools are those which give each person who uses them the greatest opportunity to enrich the environment with the fruits of his or her vision" [21].

Jonas started his "Search of an Ethics for the Technological Age" – as the subtitle of his book reads [1984] – with a critique of Kant's categorical imperative "Act so that you can will that the maxim of your action be made the principle of a universal law" [10]. The addressee of the categorical imperative is the individual in his or her private conduct and shall ensure that person's own self-determination. The new imperative of responsibility of Jonas encompasses it all – the technological issue, the natural issue, and the social issue. It formulates clearly: "Act so that the effects of your action are compatible with the permanence of genuine human life", which is equivalent to "Do not compromise the conditions for an indefinite continuation of humanity on earth" [11]. That imperative hits the mark: the addressee is public policy, the collective whole. The imperative of responsibility thus covers the full range of social acts: waging wars with nuclear arms were a don't; ecological social acts such as disturbing ecological balances and undermining social metabolism were a don't; and technological ecological acts such as producing digital tools that disable human autonomy were a don't. The imperative is valid wherever – and because – the integrity of humanity is touched.

Jonas set his work – in German "Prinzip Verantwortung" (first edition [1979]) – against Bloch's *magnum opus* – in German "Prinzip Hoffnung" (written in exile in the United States between 1938 and 1947, published in the German Democratic Republic starting in 1954, in the Federal Republic of Germany in 1959 [1985] and in the US in [1986]). However, by belittling utopias, Jonas' imperative focusses on the dystopic don'ts and foregoes the opportunity to formulate a vision that is open for the participation of anybody in devising the way out of the existential threats.

The precautionary principle is apt when it can meet the condition that there are good reasons to assume that the probability of worst-case scenarios is higher than the probability of good prognoses (because irreversible tipping points might otherwise be crossed for the bad).[k] But what about tipping points for the good? The Austrian complex systems scientist Stefan Thurner emphasises distinguishing tipping points that should be deferred to increase the stability of social systems from beneficial tipping points that should be sought to trigger the change of social systems in the right direction [Thurner 2020, 248-251]. The identification of the latter tipping points is crucial, too. That would be necessary to put the techno-eco-social transformation on the right path. The end of that path is, of course, a utopia. The Great Bifurcation suggests a utopia: from a critical viewpoint, that utopia is humanity as a systemic subject of its own.

Accordingly, a planetary ethos can be anticipated by reformulating the above-mentioned imperatives with a positive emphasis on end-directedness. The transformationist imperative is:

"Act so as to contribute to the establishment and maintenance of humanity as an autonomous social system *sui generis*, endowed with self-consciousness, including conscience, in order to solve the developmental crisis and continue social evolution with a next step of humanisation."

This imperative guides the preparatory transformation of the existing interdependent social systems into informational agents. Those agents are aware of, and actively promote, their role as proto-elements for an upcoming metasystem. They are also aware of their subsequent transformation into true new subsystems of the suprasystem as envisioned in the concept of a GSIS. This imperative therefore goes beyond the approval of what the problem is and how it can be solved, but also becomes a part and start of the solution – the build-up of a planetary ethos.

---

[k] Again, the COVID-19 crisis teaches us that a non-intervention position enables the SARS-CoV-2 virus to reproduce itself exponentially, thus crossing tipping points that are bad for human lives.

3.2.1.2 *Ethical global governance*

Such a planetary ethos is the *sine qua non* for global governance. The GSIS concept does not describe details of the meta-/suprasystem beyond its basic requirements. Although it underlines the importance of an institution that implements the new planetary social relations, it by no means plans a global government that would be prone to the preponderance of national interests. The new supranational governance would not be compatible with a unipolar structure imposed by a superpower. Multipolarity options promise steps in the right direction. Ultimately, no actors must be excluded from participating in governance. Priority would be given to concepts of international law such as peaceful co-existence, collective security, no-first-use of nuclear weapons, non-interference in internal affairs and the like. Political and ethical concepts such as Mikhail Gorbachev's "New (Political) Thinking" of world politics in the face of exterminism (see the discussion in [Haug 1989, 49-126]) needs to be recalled. The thousands-of-years old Chinese "Tianxia" [Zhao 2019; 2020] should be rediscovered. Theologian Hans Küng's "New World Ethic" [1997] needs to be revisited. Calls to actions such as the 17 Sustainability Development Goals of the United Nations deserve reconsideration.

The problem lies not in the normative constant of social information evolution: it would be open to the revolutionary leap of a global universalisation to a planetary ethos.[1] The problem is that such an ethos

[1] "Despite some literature based on biologistic biases unable to imagine a transgression of the conceptual framework of the nation-state "we", transnational relations have been taking shape. There is empirical evidence of co-operation between culturally homogeneous groups several tens of thousands of years ago, between cities around five thousand years ago, and between modern states since the seventeenth century [Messner and Weinlich 2016; Neumann 2016; Grimalda 2016]. This co-operation between collective actors like groups, cities and states has already been paving the way for co-operation among the whole of humankind in the same way that dyadic, inter-personal co-operation between individual actors opened up the space of possibilities for triadic, societal co-operation. Examples are, as **top-down** models, a diversity of historical empires and contemporary regional federations with an economic or political focus like the EU as well as a diversity of

does not yet exist and needs to be imagined. Importantly, if actors act as if the planetary ethos existed, they help bring it into existence; the more actors there are, the closer they come to the tipping point that will make that positive change irreversible (Figure 3.18 – the imagined is semi-transparent). The black and grey co-operative information arrows signify an entanglement of the actors' imaginations.

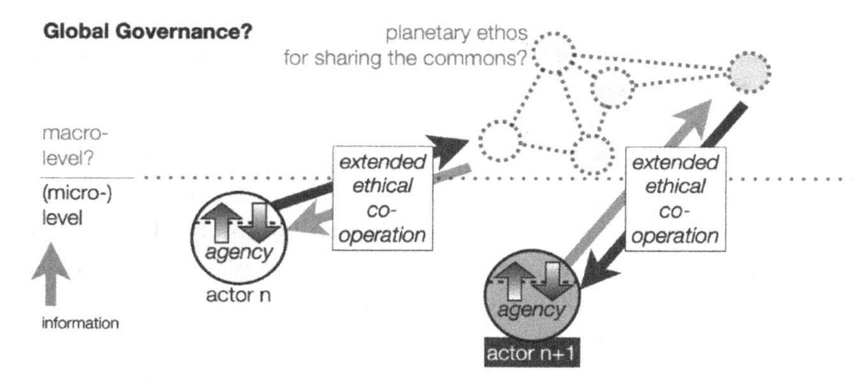

Figure 3.18. Global Governance. The preparation of a planetary ethos through as-if co-operation.

In order to become reality, Global Governance must be based on an intentional extension of the generation of co-operative information. This

---

organisations that fill the space beyond states, with the League of Nations as forerunner, and international organisations after 1945 like the UN family. Both supranational and international organisations turn rather in the direction of transnational organisations. Though they are still mirroring changing geopolitical balances of power, managers that have been running them developed an identity beyond the nation state, at a higher level [Manesse]. Besides the top-down models, another model of transnational institution building has emerged that pays attention to **bottom-up** processes too. Examples are self-regulating communities, in particular in the economic field [Djelic and Quack 2010] as well as the large number of civil society organisations (CSOs), part of which are non-governmental organisations (NGOs), in particular, international NGOs (INGOs). And there have been social movements flashing up" [Hofkirchner et al. 2019, 455-456].

must go beyond the restrictive limits to co-operation[m], i.e., the current antagonisms about sharing global commons with other social systems as well as sharing commons of any kind within their own systems. Both cases are very similar. Establishing the sharing of global commons projects the extension of restricted co-operative information generation onto a new level, i.e. one that does not yet exist. In contrast, establishing the sharing of commons other than global involves replacing the legitimation of existing antagonisms on a level that already exists. In both cases one actor can start the process and another can reciprocate. In Figure 3.18, actor n may start with a unilateral act in advance, as if the planetary ethos existed (black arrows), and actor n+1 may reciprocate in a corresponding manner (grey arrows). They do not need to imagine the planetary ethos in exactly the same way, but can accept that each partner has grounds to assume that the other partner is also inclined to the idea of planetary ethos. This constitutes an example that can be accepted by others.

The COVID-19 pandemic is once again an example for the absence of a Third. It would more than a mere act of solidarity for the rich nation-states to help the poor ones master the crisis by providing them with vaccines, waivers of patents and everything else they lack. It would be also rational to do so as quickly as possible because, otherwise, the SARS-CoV-2 viruses would have sufficient time to produce escape mutations that bounce back on those in the metropoles who have already been immunised against the older variants. A pandemic can be successfully conquered only when almost the whole world population is vaccinated. Currently, short-sighted economic interests forestall nation-states from protecting their own populations from rising death tolls. The policy focus on the intensive care capacity of hospitals does, however, at least help reduce fatalities. But members of "shadow" families, which harbour most

[m] By using the wording "to restrict" and "to extend" as opposites, I draw on an idea of Critical Psychology founder Klaus Holzkamp [1983], who makes a distinction between a "restrictive capability to act" and a "generalised capability to act". His first term means that antagonistic social relations set limits to the interaction of actors such that they restrict their reflexivity to their own. His second term means that actors enjoy together a generalised co-operation, if they succeed to strike a balance between their interests and the general interest.

vulnerable relatives such as immunosuppressed people, are exposed to contagion without sufficient protection.

3.2.2 *Remodelling social information in the age of a Great Transformation: the emergence of planetary agreeableness for global dialogue*

The considerations of social ontology under the premises of the Great Bifurcation open the possibility of an anthropocenic civilisation based on anthropocenic terraformation. Transformationist social-informational ontology investigates how the anthropocenic realities can become the content of communication among self-organising actors, in particular how agreements can be reached given the information-society mass conversation[n] (Table 3.6).

It would be an oversimplification to assume that mass-conversation as such is sufficient to initiate a successful global discourse about shaping Earth in a civilised manner. Mass conversation can even trigger intransigent public dispute on controversial positions. Reducing the globally discursive is not an option. Conversely, a global discourse in itself does not seem to suffice for a mass conversation about global concerns.

---

[n] By "mass-conversation" I mean what Manuel Castells, the Internet sociologist, has referred to as "mass self-communication". "It is mass communication because it reaches a potentially global audience through p2p networks and Internet connection. It is multimodal, as the digitization of contents and advanced social software, often based on open source programs that can be downloaded for free, allows the reformatting of almost any content in almost any form, increasingly distributed via wireless networks. *It is also self-generated in content, self-directed in emission, and self-selected in reception by many who communicate with many.* This is a new communication realm, [...] it has the potential to make possible unlimited diversity and autonomous production of most of the communication flows that construct meaning in the public mind" [Castells 2009, 70-71]. In brief, this means that although information society technologies have provided new potentialities, they are shaped by commercialisation. Thus, communication has assumed the form of rather superficial conversations, a form that has mesmerised the masses in all societies and is meeting self-centred needs.

Projection onto the mass-conversational is also not an option. None of the conflations is conducive.

The opposition to conflation would separate the mass-conversational from the globally discursive. That approach is doomed to failure. The solution is to reveal the genesis of their mutual influences in order to understand their relatedness.

Table 3.6. The consideration of non-being and being in transformationist ontology. The (Great-Bifurcation) mass-conversational and the globally/planetary agreeable discursive.

		non-being	**being**
	reduction: individualistic fallacy	**the mass-conversational:** sufficient condition for the globally discursive	**the globally discursive:** resultant of the global mass-conversational
con-flation	**projection: sociologistic fallacy**	**the mass-conversational:** resultant of the globally discursive	**the globally discursive:** sufficient condition for the mass-conversational
disconnection: anarchist fallacy		**mass-conversational**	**globally discursive**
		independent existents	
combination: transformationist ontology for a Great Transformation		**the Great-Bifurcation mass-conversational:** necessary condition for a planetary agreeable discursive	**the planetary agreeable discursive:** an emergent from the Great-Bifurcation mass-conversational

An agreeable discourse on the planetary content requires a mass conversation that bears in mind the circumstances of the Great Bifurcation. From that basis, the mass conversation can trigger the emergence of agreements through the planetary discourse. That combination is the ontological solution to how the conversational potential can be transformed into actual agreements during the discourse. A Great Transformation needs to be prepared by a transformation in communication.

Communication involves the clash of the many consciences of the multitude of actors. Convivialism provides principles that ideate the social self-organisation of actors within their structures. The Second Convivialist Manifesto adds a third element to the principles of common naturality and common humanity – the principle of common sociality:

> Human beings are social beings for whom the greatest wealth is the richness of the concrete relationships they maintain among themselves within associations, societies, or communities of varying size and nature. [Convivialist International 2020]

The manifesto goes on to conclude with the principle of legitimate individuation, which refers to limitations of individual freedom:

> In accordance with these first three principles, legitimate is the policy that allows each individual to develop their individuality to the fullest by developing his or her capacities, power to be and act, without harming that of others, with a view toward equal freedom. Unlike individualism, where the individual cares only for oneself, thus leading to the struggle of all against all, the principle of legitimate individuation recognizes only the value of individuals who affirm their singularity in respect for their interdependence with others and with nature. [Convivialist International 2020]

According to a Critical Information Society Theory, society itself represents, with its social relations, the Third. That Third reconciles individual freedom with responsibility for the commonality. In light of the Great Bifurcation, the balance between freedom and responsibility is shifting onto a qualitatively higher level. Individual freedom is increasing through the range of interactions with ever new members of all humanity, through the horizon of accessible nature widening to whole planet Earth, and through the ever more sophisticated means of mediating those activities. And, through those very activities, responsibility is also increasing.

The fifth, and last, principle of convivialism is the principle of creative opposition:

> Because everyone is called upon to express their singular individuality, it is normal for humans to be in opposition with each other. But it is only legitimate for them to do so as long as this does not endanger the framework of common humanity, common sociality, and common naturality that makes rivalry fertile and not destructive. Politics inspired by convivialism is therefore politics that allows human beings to differentiate themselves by engaging in peaceful and deliberative rivalry for the common good. The same is true of ethics. [Convivialist International 2020]

That principle is key to the conduct of communication.

3.2.2.1 *The imperative of a planetary agreeableness*

Similar to Figure 3.17, Figure 3.19 illustrates a possible extension of the second stage of the discourse ability to an omniadic, planetary agreeability.

The x-axis shows phases of interpersonal, societal and global compatibility of communication, referring to the reach that the symmetry of agreements has. Zhao argues:

> Universal values are generally understood as values applied to every individual, an inherent error bound to result in the following paradox: if a certain culture can believe its values to be applicable to everybody, then every culture can believe the same, thus resulting in conflicts among civilizations. [...] Compatible universalism, on the other hand, considers universal values as those applied to every interrelation; namely it anchors universal values on symmetrical relations rather than on unilateral individuals, thus avoiding the paradox in values. The basic principle for compatible universalism can be stated as follows: any value that can be defined by symmetrical relations is a universal value. [...] Any value that cannot be defined by symmetrical relations only represents personal preferences or the specific values of a particular group. [Zhao 2019, 60]

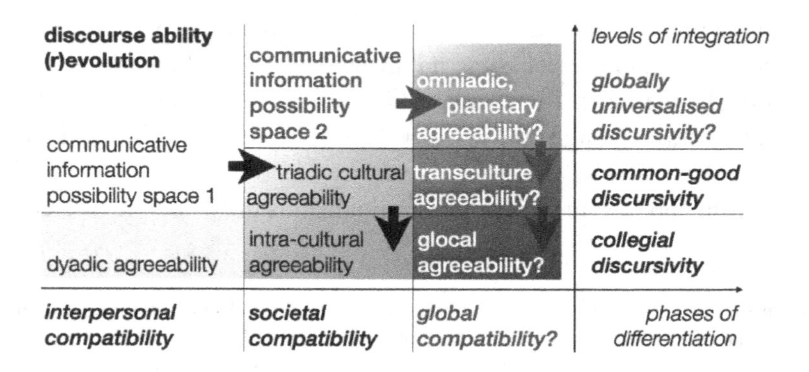

Figure 3.19. Discourse ability (r)evolution. Leap to planetary agreeability?

Accordingly, compatibility can be measured by the symmetry an agreement displays. The more symmetrical it is, the more agreeable it is. Deliberation, discussions and negotiations should be undertaken by all with the aim of symmetry in mind.

The y-axis indicates the levels of collegial, of common-good and of globally universalised discursivity. Collegiality corresponds to discourses on the lowermost level, the *commune bonum* to the next-higher level of triadic discourses, and the global universalisation to the discourse on the top-level, which is yet to be realised.

3.2.2.2 *Agreeable global dialogue*

Planetary agreeableness is the precondition for communication processes and products to master the global challenges, discuss the root causes of the Great Bifurcation, and settle on a course of action to implement a Great Transformation. Such a discourse is to be held on every aspect of necessary systemic changes in social, environmental and technological concerns and to be held with the participation of as many actors as possible. This is the Global Dialogue. The Global Dialogue is the basis of Global Governance. Global Governance cannot be reached without Global Dialogue.

Agreeableness does not mean the property of aiming at compromises that water down that which is ineluctable, inevitable, inalienable. It means that actors are partners in the same dialogue such that the same conditions apply to all of them. The search is for agreements that do not favour one partner or disrespect another partner, but that bolster their ethical co-operation as if a planetary ethos for sharing the commons existed (Figure 3.20). The actors conduct the dialogue as if a planetary ethos existed. Therefore, actors need to extend the limits of antagonistically restricted communication. As with co-operation, at least one actor, say, actor n, transcends the limits of restricted communication. Actor n configures an expression, refers to a content and appeals to a context, all of which address an imagined Third and the will to implement it (black dotted arrow). Actor n individually shapes the contribution to some collective action so as to make it compatible with possible individual contributions of the partner, say, actor n+1, under the premise that another kind of social relations is concurrently established. The same holds true for the partner. Actor n+1 may also be instigated or reinforced to transcend the limits of restricted communication and may respond to the sender to interact accordingly with reference to the own imagination of a Third (grey dotted arrow). The iteration of their extended communication may raise the compatibility of their positions. Even opponents can convert into proponents and support a common proposal in view of a Third to be commonly construed [Hofkirchner 2015, 109]. Having done so, both partners of the Global Dialogue have become proponents of Global Governance in that their extended agreeable communication bolsters their ethical co-operation.

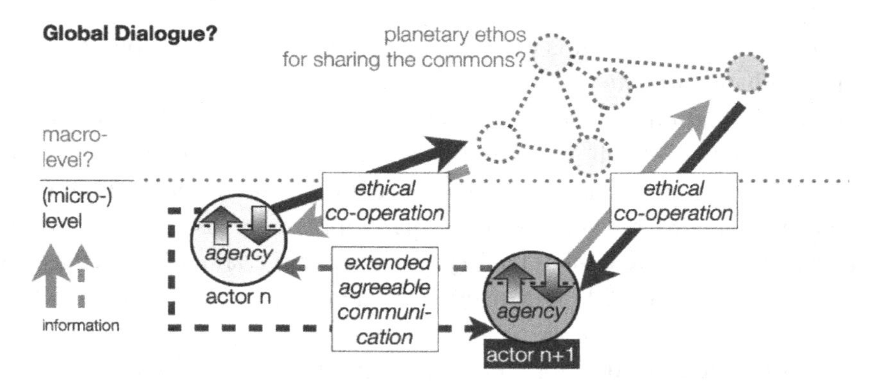

Figure 3.20. Global Dialogue. The bolstering of planetary ethical as-if co-operation through as-if communication.

Agreements must be substantiated solely by facts, not by fakes. The dialogue must be conducted truthfully. Science needs trust. Importantly, debates that focus on emotions cannot cope with conflicts because emotions cannot decide on what is true. According to the German dramaturge Bernd Stegemann, who is working on a new debating culture, conflicts can be settled recognising a common fundament based on a shared reality that can make different interests compatible [Stegemann 2021, 111]. Public communication on conflicts is not about the feelings of each other. That would merely create the false appearance of a personal, private, intimate communication [185]. What is needed is thinking in complex relations, not any reduction to individual indignation [191].

The SARS-CoV-2 pandemic is once again an example for the failure of such properties, at least with regard to German-speaking countries. Right-wing political parties started campaigns against the government measures and successfully recruited a considerable minority of the various countries' populations to refuse vaccination. This was achieved by obvious fake news while at the same time blaming, in contradistinction, the governments for spreading fake news. It is true that those governments have not been very decisive in arguing for vaccination. That minority is probably driven by prejudices against so-called conventional medicine and predilections for so-called alternative medicine in public health. This is partly a heritage of vaccination scepticism over the past 150 years, but may also reflect the Nazi era, which had crusaded against Jewish doctors and, in turn, had propagated Rudolf Steiner's "Anthroposophy" and similar notions, thus reinforcing esotericism as an anti-scientific worldview.[o] Sceptics of vaccination generally have a poor knowledge of the topic, much gleaned from social networking sites. After the fourth infection wave and the frustration of insufficient immunisation, the governments felt forced to consider compulsory vaccination and partial lockdowns for the unvaccinated only. The intransigence of the sceptics prohibits an agreement. When appeals to individual responsibility fail, the state is prompted to enforcement for the general interest. The position of the sceptics is not compatible with agreeable communication because it causes fatalities. It revives the social Darwinist version of the "survival of the fittest".

3.2.3 *Reframing social information in the age of a Great Transformation: the emergence of planetary mindfulness for global citizenship*

Conceptual reflexivity is a constant facing new conditions, given the requirements of pan-humanism, alliance with nature and digital humanism

---

[o] It therefore comes as little surprise that vaccination sceptics are also found among the medical fraternity. On 14 December 2021, 199 doctors published an open letter in which they demanded the withdrawal of the president of the Austrian medical association on grounds of "anti-scientific propaganda" in favour of the vaccination. The Austrian medical association counts about 47,000 doctors. The letter's argument is that vaccination is more dangerous than infection. Suffice it to say that that contradicts the evidence of almost all studies on the respective morbidity and mortality rates.

for social relations. How can it develop further to respond to these requirements? How must cognition be shaped to harmonise with agreeable communication and conscientious co-operation?

Those questions boil down to how the imaginary of the multi-crisis can surmount superficiality and yield the integral global reflexion required for a Great Transformation (Figure 3.7).

Table 3.7. The consideration of the apparent and the essential in transformationist epistemology. The (Great-Bifurcation) imaginary and the globally/planetary mindful reflexive.

		apparent	**essential**
con-flation	**reduction: individualistic fallacy**	**the multi-crisis-imaginary:** sufficient condition for the globally reflexive	**the globally reflexive:** resultant of the multi-crisis-imaginary
	projection: sociologistic fallacy	**the multi-crisis-imaginary:** resultant of the globally reflexive	**the globally reflexive:** sufficient condition for the multi-crisis-imaginary
disconnection: anarchist fallacy		**multi-crisis-imaginary**	**globally reflexive**
		incommensurable knowledge	
combination: transformationist epistemology for a Great Transformation		**the Great-Bifurcation imaginary:** necessary condition for a planetary mindful reflexive	**the planetary mindful reflexive:** an emergent from the Great-Bifurcation imaginary

On contrast, the individualistic fallacy supposes that the globally reflexive is a plain resultant of the empirical imaginary of the multi-crisis. This is a mechanistic framing and is untrue. The imaginary is not a sufficient condition. People can be frightened to a degree that they are unable to gather a clear thought; they can become fatalistic; they can still misunderstand the nature of the challenge and choose a wrong concept of the global situation.

Sociologism recognises that the hidden structure of global reflexivity determines how the crisis is imagined. Of course, there are feedbacks from reflexivity. Nonetheless, two aspects remain unclarified: how reflexivity becomes global, and how to do justice to independent features of the images.

Methodological anarchism, in turn, disavows sufficient or necessary conditions.

Transformationist epistemology focuses on the concept of the necessary condition. The appearance is the necessary condition, and the essence is the emergent. Here, however, appearance and essence are framed to inhere an important feature: the imaginary has to mirror the Great Bifurcation, and its reflexion must be effected by a planetary mindset. Only by that approach can the picture nudge the emergence of the mindset and can the mindset further develop the picture.

The mindset referred to here is the mindset of a relational subject introduced in section 2.2.3.3. Above all, this means that – independent of the global challenges – in Mead's [1934] distinction between "I" and "Me" "the 'I' always reflects on the 'Me'" such that "each of us is always able to become something new", as Douglas V. Porpora and co-author Wesley Shumar write [Porpora and Shumar 2010, 208].

> Whereas at least all higher animals may be considered conscious, humans alone are said to be self-conscious, precisely because of their distinct ability to regard their own selves as an object. It is this very capacity for self-consciousness moreover, that makes humans into selves in the first place [...]. We are selves, in other words because we can and do treat ourselves as selves. Indeed, [...] there cannot be any concept of self-consciousness without a self that is simultaneously the subject and the object of that consciousness [...]. Self-consciousness and selfhood thus go together. [Porpora and Shumar 2010, 206-207]

Archer [2003] categorised self-reflection and conceptualised "meta-reflexives" as having what Porpora [2001] had referred to as moral purpose, that is, individuals "who care not so much about instrumental matters [as autonomous reflexives do] but rather inhabit a moral universe and reflect deeply on ethical matters" [Porpora and Shumar 2010, 209].

In the age of the Great Bifurcation, relationality of the subjects means that reflexivity includes their concern of being part of an awakening humanity. They can conceptually grasp that humanity as a supra-individual subject and can help give birth to it by establishing a planetary ethos.

Is there an outstanding feature characteristic of a planetary ethos? Convivialism names one. The Second Convivialist Manifesto – after claiming the principle of common naturality, the principle of common humanity, the principle of common sociality, the principle of legitimate individuation and the principle of creative opposition – claims an imperative that represents the core of a planetary ethos and of planetary mindfulness. That is the imperative of hubris control:

> The first condition for rivalry to serve the common good is that it be devoid of desire for omnipotence, excess, *hubris* (and a fortiori *pleonexia*, the desire to possess ever more). On this condition, it becomes rivalry to cooperate better. This principle of *hubris* control is in fact a metaprinciple, the principle of principles. It permeates all the others and is intended to serve as a regulator and safeguard for them. For each principle, pushed to its extreme and not tempered by others, risks being reversed into its opposite: the love of nature or that of abstract humanity in hatred of concrete men; the common sociality in corporatism, clientelism, nationalism, or racism; individuation in individualism indifferent to others; the creative opposition in the struggle of egos, in the narcissism of the small difference, in destructive conflicts. [Convivialist International 2020]

This is that mindset of self-limitation that Illich [1973; 1975] had in mind.

3.2.3.1 *The imperative of a planetary mindfulness*

A revolution in reflexivity will continue human cognitive evolution (Figure 3.21).

The figure follows the underlying idea behind Figures 3.19 and 3.17. Two stages of evolution are sketched to envision a revolution to a third stage. Here, evolving mindfulness is tackled from the dyads over the triads to a possible omniad of humanity.

The x-axis shows phases and differentiates relationality, a term introduced above and also presented by Zhao [2019], as one key concept among others for a new Tianxia system. The differentiation is, as always, between interpersonality, sociability and, finally, globality.

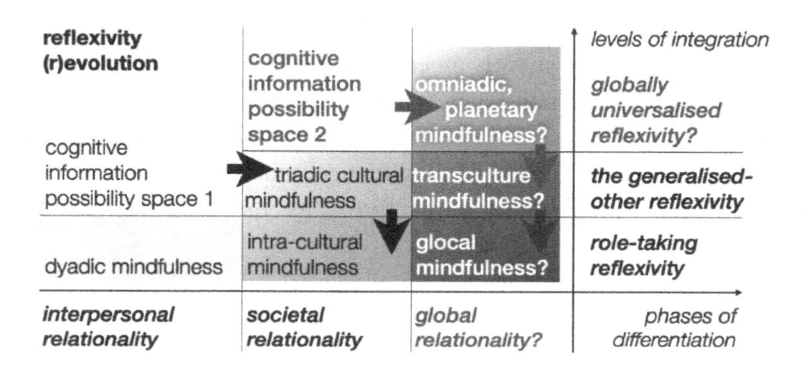

Figure 3.21. Reflexivity (r)evolution. Leap to planetary mindfulness?

The y-axis shows levels and integrates role-taking reflexivity into the generalised-other reflexivity. That, in turn, might ultimately be integrated into a globally universalised reflexivity. "Role-taking" is a label for the reciprocal, empathic capability of conceptualising what is called a "theory of mind" or an understanding of mental states as conditions for behaviour. The "generalised other" [Mead 1934] refers to the ability of generalising encounters with different others to reach a level of understanding of society as a moral community. That generalisability can be extended to a possible global level of reflexivity.

Relational reflexivity of any kind presupposes the capability to distance oneself from one's self. Only then does it become comprehensible that contingencies might have influenced the interests of one's own being as well as the otherness of interests of other human beings. Only then does it become conceivable how, from a bird's eye view – the view from the Third – the social systems relations can be shaped so as to satisfy all sides. In that context, it makes no sense to defend a position of "never forget, and never forgive". That stance forestalls the opportunity of the "I"s to renew themselves in concert with the new "Me"s demanded from society.

In summary, what humanity has learned in the two stages so far and what has been reflected in the levelling of reflexivity can, in principle, be extended to a new stage.

3.2.3.2 *Mindful global citizenship*

Margaret Archer's "meta-reflexives" are the prototypes of global citizens. Their conscientiousness merely needs to be lifted up through levels of actors from the individual level via any intermediate level of social systems to the global level. This would form a global conscience and make them global citizens. Global Dialogue necessitates global citizens as mindful actors, Global Citizenship forms as its necessary condition (Figure 3.22).

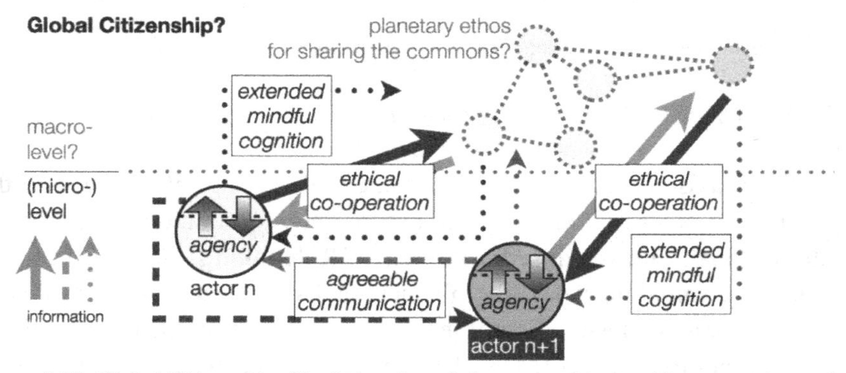

Figure 3.22. Global Citizenship. The bolstering of planetary ethical as-if co-operation and as-if communication through as-if cognition.

In addition to extending information capacitation to ethical co-operation and agreeable communication, the generation of cognitive information – which is currently restricted because of antagonisms concerning the (planetary) commons – can be extended to envision planetary ethos. Such cognitive information can put mindful cognition in practice as if a planetary ethos existed, it can bolster agreeable communication as if a planetary ethos existed, and it can bolster ethical co-operation as if a planetary ethos existed.

In the figure, both actor n and actor n+1 carry out that extension.

> At least one actor transcends the limits of her restricted cognition and extends it to include reflexivity about the social relations she and her opponent are realising in their behaviour. By adopting a critical stance, by detaching herself from the role assigned to her by the antagonistic relations, by refusing to accept the position she

holds as plain negation of the opponent's position, she can break out of the antagonistic behaviour and take unilateral action in advance. By doing so, she demonstrates that the antagonistic social relation cannot be maintained any more, since one side of the contradiction has been lost. The contradictory relation morphs into a contrary relation, the antagonistic one into an agonistic one. At least one opponent defies being irreconcilably opposed to the other, thus mutating into a proponent simply stating her position. [Hofkirchner 2015, 108]

If this quote describes the cognitive extension of actor n, then actor n+1 – through the Global Dialogue they conduct *in statu nascendi* and through the Global Governance they practice *in statu nascendi* – may be inclined to extend his or her cognition too. He or she might become a proponent, as is actor n, and then forward similar proposals. As soon as both proponents agree to let their claims be co-existent in communication and also entangle their contributions to the common good in co-operation, they recognise the new relation between them. That might result in an ultimate shift from agonism to synergism, albeit initially only between the two of them [Hofkirchner 2015, 110].

Global Citizenship, Global Dialogue and Global Governance are framed here as nested spaces such as cognition is nested by communication and communication by co-operation, one being the necessary condition for the next. Nonetheless, such a framework does not imply that change in line with the three imperatives can be proliferated only in one direction. Change can occur at any level. This framework is one of enabling spaces [Peschl and Fundneider 2012]. These spaces of possibilities are anchored in their nested holarchy, and that holarchy works as scaffolding along which real developments can emerge.

This framework enables the respect for the informational imperatives to the greatest extent without use of strict enforcement. The social space of global citizens helps attain global concerns, the social space of planetary communication furthers global dialogue, and the social space of the global public contributes to global consciousness with a global conscience.

According to that framework, global governance is distributed along nested information processes: every level provides a space for information processes that are conducive to the emergence of information processes that comply with the imperative on the next higher level and every level is a space that reinforces those information processes that it necessitates on the next lower level. The meta-reflexions taking place in the space of global citizens are conducive to the global

> dialogue and the space of the global dialogue that includes the former space shapes the reflective processes there as these are part of it; at the same time, global dialogue is conducive to global governance, while global governance demands global dialogue as part of it. Thus, the model proposed here conceptualises global governance as unfolding in time over levels of relative autonomy, as emergent product of a punctuated bottom-up process that entails a top-down process that reorganises the preconditions from which global governance arises and upon which it builds. Since individual actors reside on the bottom level, interact with each other on the intermediate level and produce social relations of synergy on the top level once they co-act, global governance is a process of social self-organisation in which agency is the driving force that is nudged by the structure it produces. Nevertheless, it is an open-ended process that scaffolds from the local to the global in a subsidiary manner. [Hofkirchner et al. 2019, 459-460]

That framework is a framework of governance from below, from dialogue and from citizens. It is contingent. Any event at any level can radiate and yield repercussions in unforeseeable ways.

Returning to COVID-19: in some countries – and thus considerably obstructing the worldwide control of the pandemic – the crisis is a lost opportunity to overcome restricted cognition. It was clear from the onset that a certain level of immunisation of a given population needs to be attained, much like for other diseases. Nonetheless, many people refuse vaccination as blunt state interference in their private domains, particularly in their bodily integrity. The restrictive understanding is that any state interference is seen as an unacceptable limitation of personal freedom. Such an understanding overlooks that this itself is in effect a blinded misunderstanding: it limits the freedom of others when the refusal to vaccinate causes others to become infected, hospitalised and potentially dying. Restrictive understanding is a lack of solidarity. In such emergency situations, when individual responsibility to act according to one's moral duty is not assumed, the state must step up to help guarantee the personal freedom of those who require solidarity. So-called conspiracy theories play an important role in cognitive restriction. This refers to wrong generalisations, those not corroborated by facts. Fake news is produced instead, then defended by hate speech that, ultimately, nourishes violence at demonstrations.[p]

---

[p] Let me substantiate the impact of restricted cognition – and the propagation thereof by political populist forces – by referring to empirical data gathered in Austria. As of

3.2.4 *The Principle of Convivialism*

Critical emergentist transformationism theoretically approaches noogenesis in the age of a Great Transformation. It is about foundations of a Critical Information Society Theory (CIST) because it enquires into the requirements of social information in order to manage a techno-eco-social transformation into a Global Sustainable Information Society (GSIS). It refers to ideas of convivialism because that movement is an initiative to bundle academic thinking to provide a scientific and inclusive participatory solution to save humanity from extinction. Imperatives of the information society have been formulated in this subchapter. This includes social-informational imperatives in the fields of co-operation/governance, communication/dialogue and cognition/citizenship mindsets. These must be respected as subjective objectives to guide all partitions of humankind into successful transitions. Equally these imperatives are objective subjects to be taken into consideration by sciences of social information when researching, accompanying and consulting civil society actors.

Figure 3.23 summarises the above by illustrating the presumed social-informational build-up in the age of a Great Transformation from Global Governance to Global Dialogue to Global Citizenship. The social-informational constants of Figure 2.9 are adapted to fit the adapted social-systemic functions:

(1) if normativity for all is to become planetary, then the consensualisation for the goals of the desired transformation require

2020/2021, 19 % of the Austrian population can be categorised as strict deniers of vaccination and 14 % are rather inclined to deny vaccination [Heinz and Ogris 2021, 32]. As of mid-December 2021, the Austria Corona Panel Project at the University of Vienna came to the result that the pool of those who are ready for vaccination plummeted to 1 %, given 75 % vaccinated with at least one dosis and yet 17 % deniers [Eberl et al. 2021]. Many of those who back the demonstrations of deniers vote for extreme right-wing parties and share anti-scientific attitudes. They believe in common sense over and above scientific studies. Many also adhere to "conspirituality" (a "hybrid system of belief" that synthesises conspiracy theory and spirituality [Ward and Voas 2011]). They also believe in homoeopathic treatment and the afterlife [Eberl and Lebernegg 2021].

that ethicalness corresponds with the social relations of a world society;

(2) if discursivity of the many is to become planetary, then the consilience on the collaboration of the transformation tasks requires that agreeableness corresponds with intersubjective world-networks of actions; and

(3) if single reflexivity is to become planetary, then the conceptual co-ordination of operations for the transformation requires that mindfulness corresponds with the subjective acts that have worldwide reach.

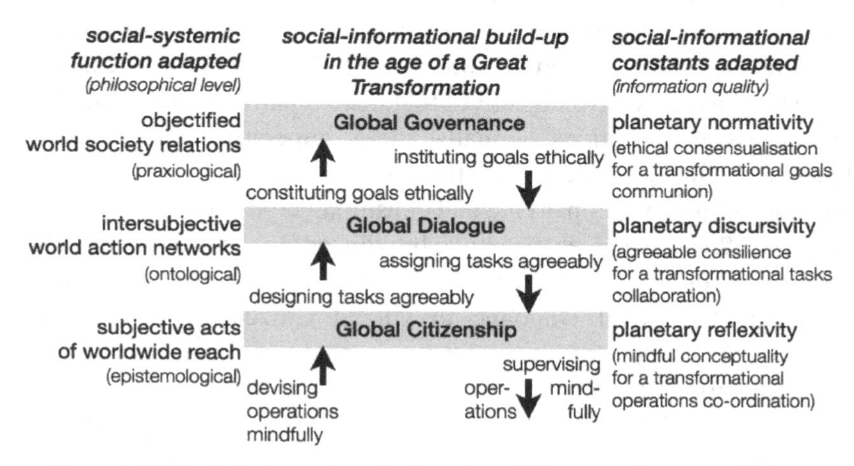

Figure 3.23. Social-informational build-up in social systems in the age of a Great Transformation.

Ethicalness, agreeableness and mindfulness are the three new convivial information qualities that form the imperatives of the build-up levels of social information for a new planetary era. These imperatives emerge by the creation of requisite social information to close the complexity gap posed by the global challenges and to guide the Great Transformation into a Global Sustainable Information Society (GSIS).

(1) The imperative of planetary ethos implies the creation of requisite information in the form of wisdom, which demands the priority application of the general interest of united humanity. This imperative is about recognising rational commitments to the common human

destiny. Such recognition needs to overcome restrictions anchored in imperial intentionality.

(2) The imperative of planetary agreeableness implies the creation of requisite knowledge diversified to shape the future by reconciling interests; this requires permanent (re-)assurance by renegotiation. This imperative is about searching for compatible solutions, which must abstain from restrictions due to intransigent self-righteousness.

(3) The imperative of planetary mindfulness implies the creation of requisite mindsets that provide an understanding of individual self-limitation, i.e. that it is best to serve one's own true interests as a member of the human race. This imperative is about relational reasoning. The faculty to judge the relations by which the individuals can become interwoven with social systems at different levels, including world society, must transcend restrictive idiotic identities.

Humanism revisited, update IV. A true humanist view needs to embrace the above imperatives. The imperatives of planetary ethos, agreeableness and mindfulness – as the convivial noogenetic conditions for a Great Transformation – are needed to update humanism to tackle the most severe challenges to humanity ever. This would enable taking the right action in a crisis that is not only man-made but can also be solved by human ingenuity. Convivialism is backed up by the planetary sociogenetic insight into the causes of the Great Bifurcation as well as by noogenetic eudaimonism and sociogenetic commonism. They characterise, in principle, the whole enterprise of anthroposociogenesis.

> The conclusion is that the current state of human evolution has been reached as emergent response to requirements of co-operation through two steps in anthroposociogenesis, namely, from the living together of individual monads towards a joint interaction in dyads and from that to a collective working together that was mediated by social relations – which are the social system's relations of the organisation of the commons – such that a triad has taken over the co-action of humans: a meta-level was constructed as a Third that relates the interaction of the group members as a Second and any action of a member as a First. Now that global conditions require global co-operation, the Third needs to be extended to another level ushering in a new phase. [...] From that perspective, the Great Bifurcation can be regarded as a problem of coming-of-age of humanity. By accomplishing that evolutionary step, the rise of co-operative organisation would enable 'the emergence of a coordinated and integrated global entity' (Stewart, 2014, 35) not seen before. [Hofkirchner 2020d, 6]

In order to accomplish this unprecedented revolutionary step, "those imperatives, investigated by social sciences and humanities, need to be provided to civil society by translational sciences, all of them integrated and implemented by the new paradigm shift as transdisciplinary basis" [Hofkirchner 2020d, 6].

Humanism is convivial, planetary, eudaimonic, commonist.

Ultimately, it can be defined:

Transformationism. *Transformationism is that critical social elaboration of* weltanschauung, *that critical conception of the social world and that critical social-scientific way of creating knowledge that devises the criticist Principle of Convivialism by applying the criticist Principle of Eudaimonism to the circumstances of the age of a Great Transformation, based upon the criticist Principle of Planetarism.*

The **Principle of Convivialism** states: there are, regardless of whether or not thematised, associated with the entry of informational social systems into the age of a Great Transformation,

(1) the emergence of planetary ethos as normativity for the common-destiny humanity;

(2) the emergence of planetary agreeability as discourse ability for the anthropocenic civilisation;

(3) the emergence of planetary mindfulness as reflexivity for anthropo-relational subjects;

such that the planetary ethos, the planetary agreeability, and the planetary mindfulness are social-informational imperatives. They thereby specify the eudaimonistic noogenetics and the planetary sociogenetics. These three imperatives are instantiated in a hierarchy of informational levels, whereby each is an emergent meta-level on its own, i.e., a Third. That Third represents an essential informational property of social self-organised systems, actualised in ideations by the interaction networks of the actors as plural Seconds, in which the participating actors – humans or other informational social systems – represent single Firsts. Any informational social system in the age of a Great Transformation is exposed to these required specifications of informational self-organisation.

Humanism is instantiated as being convivial.

Part III

Towards a Science for, about, and via the Techno-Eco-Social Transformation

Chapter 4

From Critical Utopia to Visioneering. Rethinking Future Information Technology: Emergentist Techno-Social Systemism

Tools are intrinsic to social relationships. An individual relates himself in action to his society through the use of tools that he actively masters, or by which he is passively acted upon. To the degree that he masters his tools, he can invest the world with his meaning; to the degree that he is mastered by his tools, the shape of the tool determines his own self-image. Convivial tools are those which give each person who uses them the greatest opportunity to enrich the environment with the fruits of his or her vision. Industrial tools deny this possibility to those who use them and they allow their designers to determine the meaning and expectations of others. Most tools today cannot be used in a convivial fashion.

– *Ivan Illich: Tools for Conviviality, 1973* –

We have already reached the point of starting to believe that the algorithm knows us better than we know ourselves. It then comes to be seen as a new authority to guide the self, one that knows what is good for us and what the future holds.

The return to a deterministic worldview becomes a possibility again.

– *Helga Nowotny: In AI We Trust, 2021* –

Part II elaborated theoretical foundations of a science of transformation. Chapter 2 dealt with social and social-informational transformation on a general level, as applicable to the whole adventure of anthroposociogenesis including noetic issues from a societal level down to an individual level, Chapter 3 focussed on objective and subjective conditions for the further development of socio- and noogenesis in the current age of the Great Bifurcation and a requisite Great Transformation into a Global Sustainable Information Society (GSIS) as a critical utopia. Cornerstones were laid out for a Critical Social Systems Theory (CSST) and a Critical Information Society Theory (CIST) on a high level of abstraction as well as on a level tailored for humanity's current conditions.

A Great Transformation encompasses changes in all social co-systems such as a plurality of societies and all the cultural, political and economic subsystems along with the ecological (eco-social) and the technological (techno-(eco-)social) systems. The techno-social[a] system is the most basic system of the systemic build-up of a fully-fledged social system. As such, it exhibits the most accelerated changes that materialise anyway (as opposed to systems that encapsulate such changes). Thus, it holds the key position for inducing or bolstering change on the higher levels.

Part III is devoted to discussing the role of Information Technology in support of a Great Transformation. This discussion contributes to the foundations of a science of transformation. Based upon a CSST and a CIST, it carves out building blocks of a Critical Techno-social systems Design Theory (CTDT) to help shape the future. Those building blocks rely on emergentist techno-social systemism and zero-in on Digital Humanism.

Chapter 4 departs from a critical utopia that was analysed and synthesised in Chapter 3 in order to consider implementation. This is precisely what the term "visioneering" means. Visioneering is composed of the terms vision and engineering. Kim and Oki emphasise that

> [...] vision is different from goal and objective. Vision is the documented purpose that is detailed, customized, unique, and reasonable (Munroe 2003). A goal is a general statement of intent that remains until it is achieved or no longer needed as the direction changes (Maser 1999). An objective, on the other hand, is a specific and product-oriented statement of intended accomplishment that is attainable,

---

[a] I use the term "techno-social" to better communicate that the techno-social system is a specification of a social system. This is the same intention as in other researchers' terms, e.g., Günter Ropohl's "socio-technical systems" [2012].

> observable, and measurable by specifying no more than *what*, *where*, *when* and *how*. In contrast to objective, vision focuses on *why*. Therefore, vision does not change but becomes refined, whereas plans or strategies to achieve it (e.g., goals, objectives) remain flexible and changeable. [Kim and Oki 2011, 250]

> Engineering, on the other hand, is skillful direction and creative application of experiences and scientific principles to develop processes, structures, or equipment. Consequently, visioneering requires the synergy of inspiration, conviction, action, determination, and completion (Stanley 1999). [Kim and Oki 2011, 248]

Visioneering is "the engineering of a clear vision", as Kim and Oki put it, when referring to Senge [1990] and Stanley [1999]. Visioneering requires governance to provide a vision, management to provide the operationalisation of that vision, and monitoring to provide observation-based feedback on the implementation of that vision [Kim and Oki 2011, 247-248]. It therefore differs from mere visioning, that is, imagining, as the authors illustrate with the case of sustainability science:

> Envisioning a sustainable world is an important first step toward sustainability. Without engineering it, however, the vision will not stick and just visioning a sustainable future will remain as a daydream. [Kim and Oki 2011, 250]

> According to Costanza (2003), visioneering for problem solving in social-ecological systems (SES) requires the integration of three processes: (1) vehement envisioning of how the world works and how we want it to be, (2) systematic analysis conforming to the vision, and (3) implementation appropriate to the vision. [Kim and Oki 2011, 248]

> A sustainable future will require a purpose-driven transformation of society at all scales, guided by the best foresight, with insight based on hindsight that science can provide (i.e., visioneering). [Kim and Oki 2011, 250]

Visioneering adds to a critical utopia, which is concretely and really anchored in the potentiality here and now, by showing the way to implement it.

The concept of a GSIS presented in Subchapter 3.2 is the vision. The core part of visioneering is

- considering supporting a techno-eco-social transformation by means of shaping (information) technology,
- implementing that consideration,
- examining that realisation and

- conducting the contingent revision.

This is the topic of the current chapter.

Information Technology (IT) is often called Information and Communication Technology (ICT), thereby missing the explicit mention of Computer-Supported Cooperative Work (CSCW) to follow the informationist Triple-C Model. The design, the model and the frame of the societal grip on this technology is variously approached by sociology, depending on the underlying schools. The different logics of argumentation – reduction, projection, disjunction, combination – remain the same, but the assignments to the logics will here be replaced by labels appropriate to conceptualising the connection between technology and society according to those schools. Those labels will even be tailored for each of the design, model and frame issues to yield a better characterisation (Figure 4.1).

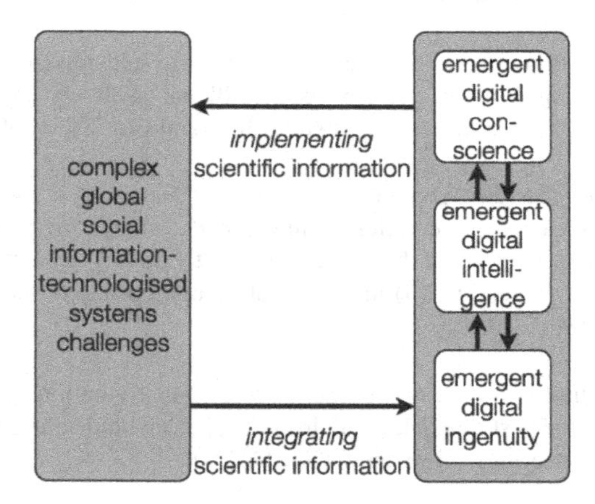

Figure 4.1. Critical emergentist, informationist techno-social systemism. The emergent digital-humanist content of the new paradigm.

Digital-humanist techno-social systemism is that theoretical step towards a science of transformation that concludes the discussion of the new paradigm. According to the new scientific and every-day thought, the complex challenges confronting global social information-technologised systems are resolvable by the emergence of digital conscience, digital intelligence and digital ingenuity. Accordingly, human conscience, human intelligence and human ingenuity become assisted by digital technologies and thereby gain a new quality in the form of techno-social systems.

4.1 Redesigning Information Technology for the Future: the Emergence of Digital Conscience

After Hofkirchner [2010a], four approaches can be distinguished:

> • The reductionist way of thinking dominates positivistic sociological approaches. Technology is deemed to be the independent variable and society the dependent one. This is called the push approach. Technological development pushes the development of society. This approach of sociology of technology is known as 'techno-determinism'.
>
> • Projective thinking is characteristic of interpretivist schools in sociology. Society plays the role not only of a necessary but also of the sufficient condition that results in technological constructs. Society pulls, so to say, technological developments. That is 'social constructivism'.
>
> • That technology and society are incommensurable fields, which are disjunctive, is a statement that is part of postmodernist strands of thought, here called 'techno/social dualism'.
>
> • Critical thinking sees a dialectic of technology and society. There is a circular influence, the first shapes the second and the second shapes the first. It is a 'mutual-shaping' approach as represented in integrated technology assessment and designing of technology. [Hofkirchner 2015, 100-101]

The last but one category is called techno-/social compartmentalism because it segregates into independent praxiological compartments that the conflationist categories try to assimilate. Therefore, the label "dualism" is henceforth used solely for the ontological assignment. The last category is then termed integrativism to signify the sublation of segregation and assimilation of the preceding categories. Integrativism is part of techno-social systemism. The term "mutual shaping" erroneously leads to co-evolutionary concepts, which overlooks the asymmetry of the cycle in question.

"Is and Ought" in IT is captured by a tension between informatisation and informedness. Informatisation was introduced in 1977 as a term to signify the process of spreading IT throughout society in a report to the president of France [Nora and Minc 1978]. Informatisation was deemed to make society increasingly responsive to information. The authors did not specify what makes up informedness, that is, which kind of information is generated, conveyed and used by IT, what is the content of information, and which purpose is it supposed to serve.

Twenty years later, an EC High-Level Expert Group on Social and Societal Aspects of Information Society was chaired by Luc Soete and included well-known scholars such as Manuel Castells, who ranks today among the most cited authors on information society matters. They finalised a report under the title "Building the European information society for us all". The basic point of departure is the tenet that "the information society signals more than a major change in the technological paradigm that underpins our society." The policy challenges ICTs raise "transcend the simplistic notions of rapid adjustment to an externally, technologically determined future in which people have little or no say" and "the sooner these are addressed the better" [European Commission 1997, 63]. The notions "wisdom" and "wise society" were introduced in this context and appeared for the first time and, unfortunately, so far, for the last time in an official document of the European Commission:

> One of the main effects of the new ICTs has been to speed up and cut the cost of storing and transmitting information a billion-fold, thereby 'energising', in the words of the Bangemann report, 'every economic sector' ("Europe and the Global Information Society", Brussels, 1994). However, these new technologies have had no such effect on the generation or acquisition of knowledge, still less on wisdom [Which we identify as 'distilled' knowledge derived from experience of life, as well as from the natural and social sciences and from ethics and philosophy.]. One would hope, of course, that society would be shifting more and more towards a 'wise society', where scientifically supported data, information and knowledge would increasingly be used to make informed decisions to improve the quality of all aspects of life. Such wisdom would help to form a society that is environmentally sustainable, that takes the well-being of all its members into consideration and that values the social and cultural aspects of life as much as the material and economic. Our hope is that the emerging information society will develop in such a way as to advance this vision of wisdom. [European Commission 1997, 16]

The adoption of the competitive Lisbon strategy rendered the vision of this report obsolete:

> The positive aspect one might be inclined to ascribe to this report is that it anticipated or, at least, accompanied the shift in European Union policy thinking from technological issues exclusively to the inclusion of economic issues testified by the subsequently accorded framework programmes for European research and development. New buzzwords – the "knowledge-based economy" and the "knowledge society" – began to partly complement and partly replace the precedent talk of the "information society". However, the turn, if any, seems to have come to a halt half way. Deep changes that affect the quality of life, environmental sustainability, individual well-being, social and cultural needs as demanded by the report are still waiting for implementation. Neoliberal worshipping seems to have been sacrificing wisdom needed more than ever. [Hofkirchner 2011, 435]

A financial crisis followed the report a little more than a decade later, entailing a global economic crisis whose impacts are still reverberating today. That was the opportune moment to question the neoliberal capitalist system and its faith in the free market. British and German Social Democrats were invited to debate on "Building the Good Society" in Europe and, in the academic field, the quest for a "good society" became topical in the emerging field of "ICTs and Society". In that vein, the director of the European division of the International Association of Computing and Philosophy, Philip Brey, held a talk at the Seventh European Conference on Computing and Philosophy 2009 titled "The Proper Role of Information Technology in a Good Society". The lifework of Swedish academic Gunilla Bradley has been intrinsically motivated by safeguarding human well-being and the search for societal conditions that enable individual self-fulfillment. This is crucial given the rapid development and deployment of converging computer, tele- and media technologies and makes her a pioneer of the good society. Her vision is "the good ICT society" [2006, 197], "the good society for all (GSA)" [2006, 229]. At the anniversary ceremony for her 70th birthday at the Linnaeus University in 2010, a volume of research essays by more than forty scholars was presented in her honour. It addresses both the normative and descriptive aspects of ICTs in human and social contexts, such as individuals, groups, organisations and their management [Haftor and Mirijamdotter 2011]. In February 2018, Atomium, the European Institute for Science, Media and Democracy, initiated AI4People to lay the foundations for a "Good AI Society". Its Scientific Committee comprised twelve experts and was chaired by well-known philosopher Luciano Floridi. In November 2018, they published their ethical framework [Floridi et al. 2018].

The author of the present book and his working group at the Salzburg University became known for their "Salzburg Approach" towards ICTs and Society [Hofkirchner et al. 2007].[b] That approach specified the good society as a Global Sustainable Information Society. The concept of GSIS made it clear that information society goes beyond a descriptive concept that is determined by informatisation. It also represents a normative concept that qualifies the informationality as that which is required to run a good society for all. Informationalisation could, in the wake of its notion for complex systems, become the term to decipher the process of capacitating society with requisite information through appropriate self-organisation (see section 3.1.2.2). The current hype about digitalisation with its focus on so-called "autonomous and intelligent systems" is merely the linear continuation of informatisation with ever new buzzwords. Unfortunately, this camouflages economic and political conflicts within and between societies and diverts attention from the proper task of coping with the global challenges to humanity.

Informatisation refers then to the Is, and Informationalisation to the Ought. Informatisiation can be viewed as the process of laying the material connections of a "Global Brain". Numerous academics have referred back to the Jesuit priest and palaeontologist Pierre Teilhard de Chardin, including politicians such as Al Gore or activists like John Perry Barlow. Teilhard was the spiritualist counterpart of Vernadsky, and his notion of the Noosphere was interpreted by followers as the building of a thinking membrane covering the planet, a globe clothing itself with a brain. Teilhard envisaged the formation of the "astonishing system of land, sea and air routes, postal connections, wires, cables, and vibrations in the ether, which cover the face of the earth more each day" as the "creation of

[b] Poe Yu-ze Wan from the National Sun Yat-sen University in Taiwan reviews our position extensively in his book "Reframing the Social – Emergentist Systemism and Social Theory" [2011] as an example of the work of critical social systems theorists.

a real nervous system of humanity, the processing of a common consciousness, the linking of the mass of mankind", as he wrote on May 6, 1925 [Teilhard de Chardin 1961; 1964; English quotes after Fleissner and Hofkirchner 1998, 205]. "In principle, this process does not differ from the evolution of primitive nervous systems into advanced mammalian brains", wrote Stonier, the author of an important information trilogy and British researcher in the journal Science and Technology. "We are now dealing with the very end of the known spectrum of intelligence" [Stonier 1992, 105]. Lévy [1997] speaks of the "technological infrastructure of the collective brain, of the hypercortex of living communities". Barlow concluded: "With cyberspace, we are, in effect, hard-wiring the collective consciousness" [cited in Kreisberg 1995]. However, a Global Brain is not the same as a global "consciousness", a global "conscience", or a global "mind" – which are rather the process and product of informationalisation (Table 4.1).

Table 4.1. The consideration of Is and Ought in techno-social systemist praxiology. The (value-based) global-brain informatising and the informed/informationalised global-conscience societal.

<table>
<tr><th colspan="2"></th><th>is</th><th>ought</th></tr>
<tr><td rowspan="2">con-flation: assimi-lative fallacy</td><td>reduction: techno-determinist fallacy</td><td>the global-brain informatising: sufficient condition for the informed societal</td><td>the informed societal: resultant of the global-brain informatising</td></tr>
<tr><td>projection: social-con-structivist fallacy</td><td>the global-brain informatising: resultant of the informed societal</td><td>the informed societal: sufficient condition for the global-brain informatising</td></tr>
<tr><td colspan="2" rowspan="2">disconnection: techno/social segregative fallacy</td><td>global-brain informatising</td><td>informed societal</td></tr>
<tr><td colspan="2">disparate takes</td></tr>
<tr><td colspan="2">combination: techno-social integrative praxi-ology for a Great Transformation</td><td>the value-based informatising: necessary condition for an informationalised global conscientious</td><td>the informationalised global conscientious: an emergent from the value-based informatising</td></tr>
</table>

The techno-determinist fallacy consists solely in the contention that informatising for a global brain would result in an informedness of society that exhibits a new quality. Accordingly, the connection of humans via their computers to the internet would reach a state of interconnectivity comparable to that of human brain cells. This is what Peter Russell argued. This magic number of nodes would be in the order of magnitude of 10^{10} [Russell 1983]. "The metaphor of the 'global brain' seems to automatically entail on a global scale what the human brain is supposed to be able to do on an individual scale, namely, to reason and to give rise to a mind" [Fleissner and Hofkirchner 1998, 206]. There is, however, no evidence for the numerical comparability between evolutionary processes of such different kinds as a (socially driven) biological one and a (socially driven) societal one.

The social-constructivist fallacy deems that society is sufficiently informed to construct the global brain. This is obviously not the case. Global conscience may be an anticipation, but such anticipations do not seem to drive and determine the development of IT.

The techno/social dualist fallacy is of no avail because it regards the global-brain informatising and the informed societal as disparate.

The techno-social systemist praxiology, in contrast, relates the global brain to a global conscience. The informatisation needs to be aligned by ethical considerations in order to be able to tip the informationalisation of a global conscience. The informational can so emerge from the informatising. Informatisation is subservient to informationalisation. Informatisation has to be tamed so as to be harnessed for informationalisation – the process of raising the problem-solving capacity of nascent world society to a level of wisdom that enables successfully tackling the problems that arise from societies' own development.

4.1.1 *The normative implications of information society theories in the Anthropocene*

Technodeterminism, social constructivism, techno/social dualism and techno-social systemism deserve further scrutiny. They can be categorised according to their evaluation of the connection conceived between technology and society. "If they look upon it favourably and highlight the opportunities, they are called eutopian. If they look upon it unfavourably and underline the risks, they are called dystopian" [Hofkirchner 2010a,

173].[c] A third evaluative category is known as "pro-active-ism", designating a pro-active stance of design that is neither eutopian nor dystopian but the only possible scientific activism. This distinction is, clearly, an ideal-typical one, as is the distinction of the sociological approaches (Table 4.2.a).

Table 4.2.a. Evaluative categories of technology-society relationships.

	eutopianism	**dystopianism**	**pro-active-ism**
technodeterminism (push approach)	technological progress = societal progress	technological regress = societal regress	
social constructivism (pull approach)	societal progress = technological progress	societal regress = technological regress	
techno/social compartmentalism	technological ≠ societal development		
techno-social systemic integrativism			designing the future by adapting technology to adapt society itself

Technodeterminism equates technological progress with societal progress and technological regress with societal regress, thereby levelling down the societal quality to the technological quality. Social constructivism does it the other way around: it equates societal progress with technological progress and societal regress with technological regress by levelling up the technological to the societal quality.

Techno/social dualism is not in the position to put the technological and the societal quality on a same level or on levels that can be connected. Progress or regress in the development of one realm therefore cannot determine a progressive or regressive development of the other.

Techno-social systemism treats either development as being capable of shaping the other, although in an asymmetrical way. Society shapes technology, technology impacts society, society reshapes technology so as

[c] Eutopia means a good place on Earth, whereas utopia means an imagined good place that does not (yet?) exist.

to shape the impacts to increase desirable impacts and decrease undesirable impacts. Hence, society shapes technology, and via technological impacts it shapes itself.

> Eutopian as well as dystopian varieties, as long as they are based upon determinism or indeterminism, are not consequential for practice. Either we live in the best of all worlds because technological progress automatically provides us with social progress or because the social progress provides us with the appropriate progress in technology. [...] Or we live in the worst of all worlds, as regress in one field yields regress in the other, and because it is determined to be so we cannot change it. [...] Indeterminism is not consequential either since neither development can influence the other. [...] It is only the pro-active variety that calls for action, for designing the future. [Hofkirchner 2015, 101-102]

Table 4.2.b. Evaluative categories of technology-society relationships with regard to IT in the age of a Great Transformation.

		eutopianism	**dystopianism**	**pro-active-ism**
technodeterminism (push approach)		digital modernisation	digital vulnerabilisation	
social constructivism (pull approach)		building a virtual community	Orwellisation	
	ecological	managing sustainability	computer-aided degeneration	
	economic	liberating knowledge for all	monopolisation of knowledge	
	political	empowering the people	surveillance, info-wars totalisation	
	cultural	elevating amenity	disinfotaining	
techno/social compartmentalism		baroque decoupling		
techno-social systemic integrativism				designing tools for a GSIS

Table 4.2.b details the evaluative categories with regard to IT that appeared contemporaneously with the global challenges.

The eutopian technodeterminist view identifies the push in modernising the world by digitalisation. These "digitisation theories" praise overall advantages that are "pushed" by ICTs. A typical representative is Negroponte [1995].

The dystopian twin of technodeterminsim views the push in virtualisation as leading to increasing vulnerability.

> Theories of virtual society (which, actually, is the title of a book by Bühl [1997]) [...] bemoan the loss of social reality through the em[e]rgence of the virtual world, the virtual space, cyberspace, and the simulacrum: man is becoming alienated from his fellow humans and emigrates to isolation; truth is no longer something to be intersubjectively consolidated but a plurality of subjective worlds. These "virtualisation theories" have been brought forward by French philosophers like Baudrillard (1995) and Virilio (2000). [Hofkirchner 2010a, 174]

The eutopian social-constructivist perspective views information society calling upon ("pulling") technology to build virtual communities, as Howard Rheingold envisioned in his famous book [1993]. Thus, information society theories – those that share the belief in the existence of certain social needs and interests that "pull" technology – might be called community-building theories (according to Rheingold). Community-building theories can be broken down according to the non-technological realms of society. Those realms are presented here in an ascending order of their nesting – from the ecological to the cultural:

> Theories of a society developing ICTs for the sake of managing natural resources (as to the environment), theories of knowledge society for the sake of providing access to the knowledge of the world for all, in particular, to better capacitate the people to earn their living (regarding economy), theories of a participatory society for the sake of empowerment of the people (at the political level) and, finally, theories of fun and leisure societies or the like (concerning culture) [...]. [Hofkirchner 2010a, 174]

The dystopian social-constructivist theories can be taken to be Orwellian theories. Orwellian theories are those

> [...] according to which ICTs are instrumentalised for more effectively plundering of the planet with regard to natural resources, for establishing knowledge monopolies in the economic sphere, for total surveillance by a panspectron which, in the political sphere, extends the Foucauldian panopticon to the whole bandwidth of electro-magnetic waves or, as regards the cultural sphere, for disinfotainment as Rheingold (2002) claims to have coined the term in the beginning of the 1990s [...]. [Hofkirchner 2010a, 175]

Techno/social dualism "provides a home for theories of unequal development of technology and society in the way Kaldor's (1982) Baroque hypothesis or Becker's (2002) position seem to indicate. These theories might be called 'decoupling theories'" [Hofkirchner 2010a, 175]. Mary Kaldor had described the arms race as something yielding to technological sophistications that are not needed. Jörg Becker had described information society as something that does not fulfil any wants of the people. In this view, technological and societal developments do not influence each other.

All those theories, devoid of the features of the techno-social systemic approach, are not theories in the proper sense of the word. They simply mirror the empirical world of information society. They do not lead to an understanding, they merely describe what can be observed and pretend to provide explanations by construing what amounts to conspiracy theories.

> Deterministic and indeterministic, eutopian and dystopian approaches provide one-sided normative descriptions of "mechanisms" that are supposed to usher in information society. Determinism is not strict, but sets the limits of possibilities only, that is, it gives room for the realisation of different possibilities. Eutopias and dystopias reflect the ambiguities of social reality that are due to existing antagonisms [...]. They reflect the tension between the opportunity to digitalise technology and the whole world and the risk of increasing the vulnerability of civilisation; the tension between the opportunity to manage sustainability and the risk of computer-aided colonisation of nature; the tension between the opportunity to liberate knowledge and the risk of its economic monopolisation; between the opportunity to empower the people and the risk of ubiquitisation of surveillance and information warfare between nation states; between the opportunity to enhance one's way of life and the risk of being manipulated by new media through disinfotainment. By reflecting those tensions, theories contradict each other. Or they reflect the tension between the opportunity of an unneeded sophistication (baroquisation) of ICTs and the risk of unmet essential needs of the populations of the world by describing contraries [...]. [Hofkirchner 2015, 102]

In contradistinction, techno-social systemism is intended as a theory of designing tools for a Global Sustainable Information Society.

4.1.2 *ICTs as value-based techno-social systems*

Social ontologists and critical realists use the term "mechanism" for the theoretical understanding of a dynamic of how social events come up. In doing so, they do not use it in a mechanistic sense. And neither does the emergentist systemist comprehension.

A mechanistic understanding implies a one-to-one mapping of causes and effects. Most technologies work like that. If, however, such technologies are integrated with social systems, this is not the case for the integrated techno-social system altogether. Techno-social systems exhibit circular causality, yielding emergent effects. On the one hand, society is assumed to shape technology and, on the other, technology is assumed to impact society. On one hand, shaping technology – exercised by subjects during the design process – yields emergent properties and, on the other, impacting society – exercised by subjects during the process of usage – yields emergent properties. Those emergent properties are due to the techno-social system's intrinsic social nature that allows for – and is inconceivable without – moral valuations. Values, norms and interests are from the onset implied by the processes of techno-social systems.

Technical function responsibility. This implication reflects not only the fact that whenever technology is designed, responsibility is taken over (willingly or not) for the technical functionality of the mechanism[d] designed. This responsibility revolves around the question: does the mechanism work efficaciously, that is, when used does it lead to the aim for which it is designed? Does it function technically? This is presumed to be a matter of fact.

Importantly, however, this perspective alone does not suffice. A morality of design and morality of use are also present.

> No act of design is without its moral dimension. But crucially, no act of use is without its moral dimension also. [...] The intention of the designer may be benevolent while the intention of the (ab)user may be malevolent, or vice versa. Most often, the physical artefacts themselves that represent the technology to hand are, in human terms, morally neutral. Neutral that is until an action is undertaken with them. Some forms of technology such as weapons represent the physical instantiation of doubtful morality. However, even these items can be employed for laudable aims, as has been epitomized in the more recent Hollywood movies about destruction-threatening asteroids. In a similar manner, surgical instruments are

[d] Mechanism in the mechanistic sense of the word, meaning an artificial configuration of causes so as to function based only on strict determinacy.

> created with the fundamental intention of healing, although this function is easily perverted by the abuser.
>
> Thus, what we see is that the morality of technology is an emergent dimension which derives from a confluence of both the intention and action of the community of designers and the intention and action of the community of users. As an emergent property, it is hard to index and quantify on a ratio scale and because of the purported neutrality of the large majority of created designs. It should be evident now that the purported neutrality actually derives from the under-specification of what the design can do, rather than the expressed intention of what it should do. [Hancock 2009, 156]

Social function responsibility. This refers to the responsibility taken over, willingly or not, for the social function that the designed mechanism is to support. That responsibility involves the question: does the designed mechanism improve the social function it is intended to fulfil, that is, does it promote the right social value, conform with the right social norm, serve the right interest? Does it function socially? This makes it primarily a matter of ethics.

Value dispositions, value qualities, ideal values in techno-social systems. Sarah Spiekermann is a leading expert in the field of "value-based engineering" – a term she coined and that is coded by the world's largest association of IT professionals IEEE in the recent Standard 7000. She provides a philosophical basis for the ethical alignment of IT design that converges with the concept of techno-social systems presented here. Referring back to the German phenomenologist tradition[e], she postulates a hierarchy of three "ontological" layers of value aspects: the most basic layer consists of the value dispositions of a material thing, i.e., features that afford certain valuable qualities; the next layer encompasses what she terms "value qualities" that are catered by the value dispositions; and the top layer is made up of overall ideal values that are actualised in the value qualities [Spiekermann 2021].

The techno-(eco-)social system is a subsystem of social systems[f] (Figure 4.2).

---

[e] E.g., Edmund Husserl, Max Scheler, Nicolai Hartmann.

[f] For the sake of simplicity, we abstract here from the definition of techno-social systems as techno-eco-social systems.

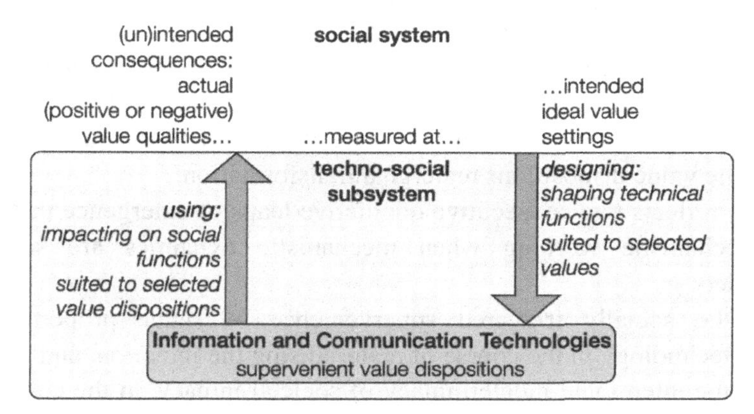

Figure 4.2. Techno-social system.

The point of departure of the design process (see downward arrow) is a setting of intended values that is set in unison with norms that apply those values and with interests that abide by such norms. These values represent the ideal that shall be realised.

The design process exerts formal causes and/or final causes in the sense of Aristotle [see Hofkirchner 2013a, section 4.1.4]. This shapes technological functions so as to support social functions that suit a selection of the ideal values. The transformation of the intended support of social functions into an appropriate shape does more than yield only properties that were expected of the designed technology. It also yields properties that go beyond the expected ones due to an almost inexhaustible number of objective affordances of the natural material that is to be subjectively processed. These properties can reinforce the intended use. They can also allow the usage for additional social functions and/or the subservience to additional values, norms and interests that can be assessed positive or negative. Finally, they can even frustrate the originally intended use. These emergent properties are value dispositions.

It is precisely these value dispositions of which a selection can be realised during the process of using that technology (see upward arrow) by exerting material and/or efficient causes. Note, however, that the impacts can concern social functions that differ from the originally intended social functions. Even more importantly, they can trigger unintended consequences: the technical functions, when entering the social fabric, can strengthen or weaken the desired social functions, influence other social functions, or bring about new social functions, all of which serve other values, norms and interests, deemed positive or negative. These emergent valuations are the actual value qualities into which the value dispositions undergo a transformation.

This reflects two consecutive qualitative leaps of emergence that rule out mechanistic working when mechanistic dynamics are socially embedded:

- the socially triggered supervenience of shape properties of technology in the course of materialising the intention, and
- the intentional indeterminacy of societal impacts in the course of putting those materialisations back in context with social functions, values, norms and interests.

Although technological means are designed with a particular intention, supervenient properties accrue in the shape of the designed technology and, irrespective of them, though based upon them, unintended consequences accrue in the impact on society. This opens a space for ethical design. Actual value qualities can be measured at the ideal value settings and improved by ongoing technology assessment and technology design. This is an ineluctable property of techno-social systems.

Informational techno-social systems. This is an ineluctable property of ICTs in the Anthropocene. Information technology as such has the potential to decrease frictions in the technical support of social functions and thereby to decrease social frictions. Social frictions are social dysfunctions: social functions no longer work for synergy, for the commons, for the flourishing of all. As they are mediated by technology, they are, at the same time, technical functions going dysfunctional, causing technical frictions. The Anthropocene started when social-systemic dysfunctions began to prevail and to endanger social systems evolution. Dysfunctionalities need to be eliminated or at least contained such that no vicious runaway is released. Responsible design must make an effort to ensure that the provision of tools for handling information processes promotes the good society, the common good, the commons. The aim must be to uncover, assemble and take advantage of favourable tipping points while dismantling disastrous tipping points. This is a key lesson to learn from the crises in the Anthropocene: the new post-crisis normalcy must differ from the old (pre-)crisis normalcy.

> The social construction of ICTs is not a 'mechanism' per se that leads to a technology that can be used for that purpose for which it is made alone. Informatisation, the penetration of society with ICTs, is not a 'mechanism' per se that leads to information society. The social system is characterised by a certain structure containing dysfunctionalities or not. The techno-social system can reinforce particular dysfunctionalities, quantitatively; it can spawn new dysfunctionalities, qualitatively; and it can support the mitigation, and even elimination, of those dysfunctionalities and the advent of new functionalities. There are contingent ways in which the 'mechanism' of mutual-shaping of ICTs and society can work. [Hofkirchner 2015, 103]

A socially aware development of technology is a must because of that contingency of paths, because the leaps are not strictly determined, and because the potentiality of technology is open for the better or the worse. As with any technology, the impact of ICTs on the social system is ambivalent. ICTs can decrease frictions in achieving synergy, but they can also increase them.

On the one hand, ICTs inhere a potential for strengthening conviviality. They can quantitatively reinforce existing or qualitatively innovate new, cognitive, communicative or co-operative social information functions – with the intention and/or consequence of serving or improving social inclusion and mitigating or eliminating existing, and preventing new, social exclusion.

On the other hand, ICTs can also be instrumented to hamper conviviality. They can quantitatively reinforce existing social information dysfunctions or qualitatively spawn new social information dysfunctions.

Well-known researchers recognise that and have striven to doing justice to the dynamic outlined here. Suffice to mention three cases:

Derrick de Kerckhove, who worked with Marshall McLuhan, stresses the potential for a watershed in the conditions of humanity. People can develop from spectators to participants because the internet and the Web are growing connections similar to a brain. Connectedness seems to be an inherent goal of self-organising processes. The challenge, however, is to make sense of the changes. Kerckhove states that people are becoming responsible for their extensions into and around the Earth [Kerckhove 1998].

Philosopher of information Luciano Floridi refers to an infosphere that he defines, in analogy to the biosphere, as constituted "by all informational entities [...], their properties, interactions, processes, and mutual relations" [Floridi 2007, 59]. ICTs are reontologising this very infosphere. "To reontologise" is another neologism Floridi coined to do justice to "a very radical form of re-engineering, one that not only designs, constructs, or structures a system […] anew, but that fundamentally transforms its intrinsic nature, that is, its ontology", which holds for ICTs in "that our technology has not only adapted to, but also educated, us as users" [Floridi 2010, 11]. Everything is becoming connected with the emergent infosphere because anything is an informational agent. The implication is that ICTs are making humanity increasingly accountable for the way the world is, will be and should be. Hence, as a common space, the infosphere must be preserved to the advantage of all. This is Floridi's message regarding his Information Ethics, developed as part of his Philosophy of Information by generalising environmental ethics onto a higher abstract level. Since technology promises opportunities and threatens with risks, it must be subjected to ethically aligned design [Floridi et al. 2018].

After Manuel Castells, who theorises in his three-volume opus [2010a; 2010b; 2011] that information society as a network society came into being through three independent levers – the needs of economy for managing flexibility, the demands for individual freedom and the micro-electronics revolution; ICTs have transformed communication into "mass self-communication" [Castells 2013, 85]. Social systems can stall creativity and rentier capitalism of the Microsoft type can obstruct the questioning of intellectual property rights. Nonetheless, the rise of mass-self-communication does enhance opportunities for social change. Powers can be challenged, so his dictum, by unveiling the construction and by revealing the exercise of power relationships.

All of these authors advocate deliberate ICTs design. And the openness for the good or evil is not an argument against, but a precondition for being able to design for the good [Hofkirchner 2010b; Dodig-Crnkovic and Hofkirchner 2011; Hofkirchner 2015, 104].

4.1.3 *ICTs enhancing cognition, communication and co-operation*

The social functions that ICTs support are human information processes. Human information processes turn up as creation of social information during

(1) cognition (one person reflects singly on his or her external or internal environment, although that reflection is nested by communicative engagement),
(2) communication (at least two persons engage jointly in sending and receiving messages to and from each other, although that engagement is nested by co-operative engagement), or
(3) co-operation (as a rule, more than two persons engage collectively in acting towards a common goal, constituting a higher-order system).

4.1.3.1 *Tools for thought*

First, the technical support of cognitive functions harbours the promise to develop "tools for thought", as Rheingold called his book in 1985, featuring what computer pioneers J.C.R. Licklider or Doug Engelbart and others envisioned [Rheingold 2000].

Mechanisation of intelligence. Prevailing trends threaten to disrupt thinking through mechanisation of intelligence and the conformation of a *Homo informaticus* along the following lines, among others (Table 4.3):

(1) Algorithmisation puts creativity under pressure. The human capacity to make generalisations and to deal with levels of abstraction is poised to get lost due by aligning with machine processing that relies solely on formal logics and mathematics. "Big Data" is an example for the abdication of intuiting hypotheses and waiting instead for the computer output to find correlations that might or might not indicate meaningful relations.
(2) Dataism focusses on gathering empirical knowledge only. An example is the neoliberal craze for measuring everything. The quantification of body performance for self-optimisation strategies is a major trend.

(3) Thinking features are being outsourced to machines to which erroneously a full-spectrum superiority is attributed. Hence the misnomer “autonomous and intelligent systems”.
(4) Knowledge is outsourced to the web, which produces artefacts. Search engines work according to the power law – the Matthew principle of “the rich get richer and the poor get poorer” – and they reinforce existing biases such as white supremacy, gender inequality, antisemitism etc.

Table 4.3. Technically supported cognitive functions. Promise and threats.

	the promise	**the threats**
technically supported cognitive functions	**"tools for thought"** (J.C.R. Licklider, Doug Engelbart et al.)	**mechanisation** of intelligence: • **algorithmisation of creativity** – unlearning human generalisations (e.g., "big data") • **dataism** – measuring everything • **outsourcing of thinking to machines** (e.g.,"autonomous" and "intelligent" "systems") • **outsourcing of knowledge to the web** – algorithms work according to the power law

Tools for convivial netizenship. What is needed in designing tools for thought in the Anthropocene is to understand that reforming thinking and education for citizens of the Earth is a *sine qua non* for any substantial change towards Homeland Earth [Morin 2012]. “What is needed is complexity thinking in every-day thinking, an understanding why trans-disciplinary approaches are required, a logic that stretches beyond deductive reasoning, systems and evolution literacy, ethical, inter-religious and inter-cultural education to build intellectual and emotional capacities of open-minded actors fit for a new planetary era” [Hofkirchner et al. 2019, 460]. Correspondingly, new tools for thought could and should be designed, e.g.,

> [...] to support initiatives in any country to reform the education systems to include pedagogics for peace, global social justice and a thriving planet, wherever applicable, from the kindergarten over the primary and secondary schools to universities and. to continuing education. Artists shall be encouraged to write fiction, to write songs, to perform theatre plays, operas, musicals, dancing, to produce pieces of

artwork, installations and exhibitions that are dedicated to the new way of thinking required or put given pieces into the context of today's challenges. Similarly, scientists should be stimulated to focus their research on such issues. [Hofkirchner et al. 2019, 460-461]

4.1.3.2 *Media*

Second, communicative functions are technically conveyed by "media".

Disinfotainment. One would assume that the media promise to facilitate mutual understanding worldwide. However, the blogosphere and what became known as today's "social media" might even disrupt it by "disinfotainment" – a word introduced by Rheingold [1993; 2002]. It might bring Morin's *Homo demens* to the centre stage with the following trends (Table 4.4):

(1) Guy Debord [1967] was the first to emphasise the role of the spectacle in society, which remains an unbroken trend. Georg Franck [2010] coined the notion of the attendance economy. This reflects the struggle for voicing oneself. Hence sensationalism, scandalisation, and rising irritability [Pörksen 2018].

(2) There is a disinformation overflow caused by manipulation and propaganda produced by the powerful in economy, politics and media. Uwe Krüger [2019] highlighted the networks of think-tanks and gate-keepers close to the elites.

(3) In the wake of postmodernism, a "post-truth" age seems to characterise the contemporary era. The capacity to discern facts from fiction and to trust in communication becomes lost for a considerable segment of the population. On social media in particular, science denialism, both-sidesism, conspiracy theories, filter bubbles, hate posts, bots etc. are becoming viral.

(4) There is also a virtual escapism, a loss of sense of reality through new means of distraction, e.g., unserious gaming, computer addiction, booming consumerism and, just recently, Mark Zuckerberg's immersive "Metaverse" and the like.

Table 4.4. Technically supported communicative functions. Promise and threats.

	the promise	the threats
technically supported communicative functions	**"media"** (Sybille Krämer)	**disinfotainment** in a divided society: • **attendance economy, the society of the spectacle** – struggle to voice oneself (e.g., rising irritability) • **disinformation overflow** – manipulation and propaganda (e.g., think tanks and gate keepers close to elites) • **"post-truth age"** – unlearning to discern facts from fiction (e.g., filter bubbles) • **virtual escapism** (e.g., "Metaverse")

Tools for convivial dialogue. The design of media in the Anthropocene needs to take the following analysis as a starting point:

> Media are influential and condition the free intercourse. It is a fact that worldwide mainstream media are biased and convey partisan interests of elites [...]. Journalists maintain not only connections to INGOs like think tanks propagating a certain political agenda but also to governments and the so-called intelligence communities of certain states. Editorial offices gather to arrange how to label certain phenomena of the political and economic world like political leaders and groups or economic measures in a way that reminds of Orwell's Newspeak. Due to deteriorating working conditions, investigative journalism is hard to practice and P.R. industries that economically outbalance media industries feed the media with fabricated news that are not questioned. Commercialisation reinforces echo chambers that trigger off the public's most primitive instincts and even diversion plays a role in that topics relevant for a peaceful future of different cultures in harmony with nature are neglected. [Hofkirchner et al. 2019, 461]

The potential for communicative democracy exists:

> Communicative spaces enable humans to grasp the world they live in through exchange with, and adapt their views to, each other. What Morin calls democracy in that context is the insight that none of us owns the absolute truth but that we can converge to consilience by adding our individual perspectives until common pictures emerge. In the age of global challenges, it is mandatory not to exclude any perspective because it might prove precious to save civilisation.

One of the most important organisational innovations deserving support by newly designed media is the establishment of platforms for transformative media outlets "that comply with the imperative of a global dialogue for the sake of civilisation" [Hofkirchner et al. 2019, 461]:

> "Transformation-oriented", "impact-oriented", "future-oriented", "solution-oriented", "constructive journalism" are denominations of a new genre. According to that, journalists shall not bring bad news but constructive news and direct their attention to problems and the attempts to solve them, including failures to learn from them. [Hofkirchner et al. 2019, 461]

Those platforms should be operated by independent bodies in order to guarantee freedom from private or political interests, which tend to block what is in the interest of whole humanity. They should also help unmask the abundant and highly varied manipulations of communication.[g]

Furthermore, those platforms could provide materials for self-organised learning and teaching materials that feed tools for thought in the sense of a pedagogics for peace, global social justice and a thriving planet.

4.1.3.3 *Technologies of co-operation*

Third, "technologies of cooperation", as Rheingold called them in a report [2005], postulate technical support for the good.

Military-informational complex. In contrast, support for the evil can be recognized when the military-industrial complex in the US – which Dwight D. Eisenhower warned of in his presidential farewell address on January 17, 1961 – is being replaced by a military-informational complex, whereby industry is being replaced by the Big Five known under the names Google, Apple, Facebook, Amazon and Microsoft. *Homo idioticus* would be established if the following tendencies are further pursued (Table 4.5):

(1) States wage cyberwars, but also private actors wage information wars.

---

[g] Albrecht Müller, who worked in German politics in the Willy Brandt administration, published a book on manipulation that, in November 2019, reached rank 2 as "Spiegel Bestseller" [2019].

(2) States strengthen their surveillance and the intelligence to control their and other states' citizens, thereby overstepping their responsibilities in that sector.
(3) The private sector is also expanding surveillance, even if with a different flavour. "Surveillance capitalism", a term Shoshana Zuboff [2019] introduced, denotes the change in the business model of the private sector in Silicon Valley. Google was the first to introduce predictive profiling of consumers.
(4) Platform capitalism, misleadingly called "sharing economies", such as Uber or Airbnb, apply dumping conditions at the cost of the common good.
(5) Digitalisation boosts the rationalisation of jobs by deploying robots and so-called cyberphysical systems.
(6) Production inflates the obsolescence of things to nourish consumerism – a baroquisation of gadgets, etc.
(7) Trans- and posthumanist, antihumanist ideologies and movements disorient the public with regard to what technical progress could be.

Table 4.5. Technically supported co-operative functions. Promise and threats.

	the promise	**the threats**
technically supported co-operative functions	**"technologies of co-operation"** (Howard Rheingold)	**military-informational complex** build-up: • **cyber/information warfare** (state/private) • **surveillance state totalisation** • **surveillance capitalism** – predictive profiling as dominant business model • **platform capitalism** – "sharing economies" against the common good (e.g., Uber, Airbnb) • **job rationalisation** (e.g., "robots", "cyberphysical systems") • **obsolescence acceleration** (e.g., gadgets) • **trans-/posthumanism** – antihumanist movements

Tools for convivial governance. In the Anthropocene, the task is to design technologies of co-operation for global governance to produce a new civilisation. This would combat the current situation in which globalisation has produced an infrastructure of a world society without a common consciousness.

> There is a growing number of social entrepreneurs, philanthropists, retired politicians, professionals, intellectuals, artists and others, working in not-for-profit sectors, who have also become part of social movements or civil society organisations, from the local to the global, all of which – individuals or collective actors – anticipate in their actions, some values, norms and principles of social relations, that could be universalised for all of humanity. They would represent the vanguard of a global conscience. More often than not, however, they are scattered around the world, focusing sometimes on a narrow section of a global challenge and become blinded through such a routine, that they lose the larger picture, if they ever had one, and hence do not develop a common, comprehensive, single integrated strategy. Many of them refrain from programmatic work, developing political demands, entering political negotiations, and even when some of them, form independent forums, or when they are invited to join international meetings or the UN system, they are sometimes not treated as being on an equal footing with the policy makers. Their influence on politics is as a consequence, rather marginal. [Hofkirchner et al. 2019, 462]

The momentum of global civil society movements and organisations that enact global ethics must be used. The idea is to innovate the social role of the UN family, e.g., by an addendum to the UN General Assembly (GA) for a politics of humanity and civilisation in the wake of Morin [2012] and to support this development by newly designed technologies of co-operation. One strategy would be to establish a permanent expert group of global civil society representatives with expertise and proved performance in transnational fields. That group would go beyond merely representing the people of the world to addressing the future population of a united world. It would be endowed with the right to elaborate on proposals on any aspect of dealing with the global challenges and to present them to the GA for further treatment [Hofkirchner et al. 2019, 462-463].

> Progress could also be achieved through states that are willing to form coalitions and implement measures, without waiting for all states to take part. Such an example is the Treaty on the Prohibition of Nuclear Weapons that was negotiated through the adoption of a mandate of the GA and signed by a group of member states. [Hofkirchner et al. 2019, 463]

The technological support of these new co-operative activities could be connected to the envisioned media platforms and form a specific focus to enhance proper media coverage. That approach would integrate tools for thought, media and technologies of co-operation. On the cognitive level this would involve online materials and online courses, video recordings of artistic performances and pieces of art, digital installations, electronic fictions books etc. that are in line with the pedagogic of a good society. On the communicative level this would involve the participation in producing and using transformative news and in events of deliberation etc. that deal with the question of which path societies should take. On the co-operative level, this would involve the processes of the UN GA addendum group as well as other organisations tasked with working out solutions for the future [Hofkirchner et al. 2019, 464].

4.1.4 *ICTs for peace, sustainability and (social) development*

Computer science has been playing a role in the Atomic Age, in climate change and in the imperial modes of production and living [Brand and Wissen 2021]. It has been contributing to research and development of digital means for thought, for media and for co-operations that have an impact on the anthropocenic situation. They can either promote peace, prevent, respond to and recover from conflicts, or be used for warfare propaganda, intelligence, communications and ICT-enabled weapons [Stauffacher et al. 2005]. They can promote creating, enabling and encouraging sustainable patterns of production and consumption and making of ICT goods and services more sustainable over their whole life cycle, mainly by reducing the energy and material flows they invoke. They can also be used to unleash their negative potential [Hilty and Aebischer 2015]. They can either promote the reduction of poverty, improve access to health and education services, and create new sources of income and employment for the poor – or be used to disrupt the status quo, leading to inequality and exclusion, both between and within countries, including industrial countries [OECD 2010; UNCTAD].

4.1.4.1 *ICTs for war or peace*

The “Manhattan Project” was the first “Big Science” endeavour. It was devoted to building the atomic bomb. Women who, prior to World War II, were used as human “computers” to calculate ballistic tables for artillery firing, were tasked with calculating other equations. They received support from IBM electric calculation machines to determine the hydrodynamics of implosions and explosions [Atomic Heritage Foundation 2017]. The end of WWII marked the tipping point of no return to a pre-atomic age; the focus turned from developing atomic weapons to improving them. The Los Alamos National Laboratory borrowed time on ENIAC (Electronic Numerical Integrator And Computer), the first Turing-complete electronic computer originally built for the US army to compute ballistic tables [LANL 2017]. The first programme run on ENIAC in 1945 was for the development of the hydrogen bomb [Nowotny 2021, 27].

During the Korean War, when the Harry S. Truman administration considered the use of nuclear weapons against China, the famous US Army General Douglas MacArthur, who “proposed measures that arguably could have triggered a Third World War”, was removed from command in 1951. The decision for nuclear attack had been removed from him and handed over to an “electric brain” that had not supported MacArthur’s approach [Anders 2016, 58]. For Austrian philosopher Günther Anders, with that step, “humanity proved that it had submitted itself to this man-made calculating robot and was willing to accept this machine as a substitute for its own conscience and acknowledge it as an oracle-machine and even as a machinic eye of providence” [59]. The story can also be interpreted differently. It seems a cunning turn of reason that the presumed subordination to a computer prevented the crossing of a tipping point that would have ushered in nuclear warfare (there was a mutual assistance pact between China and the Soviet Union that possessed 25 bombs against 450 bombs of the US [Cumings 2004]). Notwithstanding, if people were to submit responsibility to the computer, another tipping point would have been reached that would have been equally disastrous.

In the 1960s, the situation of Mutual Assured Destruction (MAD) was reached. It implied nuclear retaliation capability. The ABM (Anti-Ballistic

Missiles) Treaty was signed in 1972 between the US and the Soviet Union limited ABMs and retained the retaliatory strike option for each side. That approach made the intentional approaching of a tipping point more difficult. The Cuban missile crisis 1962 had been defused by John F. Kennedy and Nikita Khrushchev's understanding that the US could not tolerate Soviet missiles in its Caribbean neighbourhood, just as the USSR could not tolerate American missiles in its Black Sea neighbourhood.

In the 1980s, however, the US attempted to gain a first strike capability by deploying 108 Pershing II missiles and 464 Ground Launched Cruise Missiles in Western Europe. They could have decapitated the Soviet Union by targeting its military and political structures. Such a pre-emptive strike should deny the Soviet Union the ability to retaliate. Computer science provided the technological basis for establishing the threat of those high-precision weapons for surgical strikes. A tipping point would be reached due to an extremely short advance warning time for the Soviet Union. Automatisms were therefore taken into consideration. Some argued that nuclear wars could be fought and won in Europe below the threshold of an all-out war. Nonetheless, "nuclear winter" computer simulations were unable to demonstrate acceptable outcomes of war. In the aftermath, studies even questioned the conventional defense capability of European countries because of the increased civil vulnerability of industrial societies and, in particular, of information societies due to the informatisation of their infrastructure [Gonnermann and Mechtersheimer 1990; Knies et al. 1990; Roßnagel et al. 1989].

The so-called second Gulf War that liberated Kuwait by a US-led coalition of 39 countries delivered the greatest communication infrastructure initiative in the history of humankind so far. Command, Control, Communication, Computers, and Intelligence (C4I) became integrated for Information Warfare. Computer support enabled the decentralisation of the military organisation. Mission orders delegated responsibility down along the chain of command [Hofkirchner 1996]. The world public was prepared with a pro-war campaign and, once the war was started, targeted with live news broadcasts including fakes (e.g., the incubator testimony). It became a "Video Game War". Both activities were set to establish a readiness for going to war. This represents crossing a tipping point in public opinion, negating any chance for diplomatic negotiations alone to prevent or end the war.

After the demise of the USSR, President Mikhail Gorbachev had been assured by US Secretary of State James Baker that NATO would expand "not one inch eastwards" beyond the borders of a unified Germany. This was reassured by all key heads of state and government in 1990 [Savranskaya and Blanton 2017], and again reiterated to then President of the Russian Republic Boris Yeltsin by NATO Secretary General Manfred Wörner in 1991 on behalf of 13 of 16 North Atlantic Council member states [Paech 2019]. Despite this, the US State Department began to develop plans for NATO expansion in the context of strategic re-orientation whose aims went beyond mere defense [Buro and Singe 2009]. The expansion eastwards has been, among others, criticised by former Secretary of State Robert McNamara and the cold war strategist George F. Kennan, to no avail [Paech 2019].

During the NATO war of aggression against the Federal Republic of Yugoslavia, the Public Relations industry began to replace evidence-based professional journalism by offering information tailored by private P.R. firms for their own partisan agenda. An example is the training of the UÇK – The Kosovo Liberation Army – which also committed war crimes, for media performances that satisfied western tastes [Becker and Beham 2006]. This was another, privatised, way of constructing tipping points.

In the Iraq War, launched on a lie, journalists were "embedded" to work close to the military. In Syria, a special organisation, the western-funded "White Helmets", in violation of the sovereignty of that state, were in charge of propaganda footage for Islamist groups.

Vladimir Putin made years-long offers to strengthen political and economic co-operation with the EU – in particular, the idea of a common economic area from Vladivostok to Lisbon. Similar offers were made to the US, which remained unanswered. Then, the NATO member states were enlarged by the Czech Republic, Hungary and Poland in 1999 and by Bulgaria, Estonia, Latvia, Lithuania, Romania, Slovakia and Slovenia in 2004. NATO's final declaration at the 2008 Bucharest summit stated that Ukraine and Georgia were also slated to become members[h]. For the Russian Federation, this crossed a red line [Mersheimer 2014]. It also marked a missed opportunity to peacefully settle the situation. The US invested in Ukraine's Orange Revolution and, together with its European allies, tried to peel Ukraine away from the Russian (past Soviet) security sphere and incorporate it into the West. At the same time, the possibility of an association with the Commonwealth of Independent States, an alliance of former Soviet countries, was raised. These two options are seemingly inconsistent with each other but not necessarily mutually exclusive. "Trilateral talks between Kiev, Brussels and Moscow could have elevated the contradictory positions to a meta-level on which they would have become compatible and even complementary to each other" [Hofkirchner 2015, 111]. An idea of negotiating a systemic "Third" for the benefit of all of them was proposed by Russia but rejected by the EU. Thus, things turned out differently. A group not representative of the whole of the country (which is characterised by a large part of Russian-speaking people with family and work ties with Russia) seized power on 22 February 2014 by a coup carried out with the participation of fascist militia under the eyes of Western politicians. This led to the hasty secession of one part of the country and its admission to the Russian Federation, by which the Russian naval base on Crimea, leased from Ukraine, was rescued from the menace of a future NATO takeover. For the first time, the Russian side was blamed for aggression because it did not tolerate the breach of one of its vital security interests. In contrast, the US and NATO cannot claim a vital security interest in the incorporation of Ukraine. This parallels the security interests of the USSR and the US as they had been configured in the Cuban missile crisis [Paech 2014; Matlock 2021]. At that time, the US possessed around 800 foreign military bases with nearly 250,000 troops in around 160 countries [Vine 2015].

---

[h] Germany and France were hesitant. They feared deterioration of relations to Russia. Willy Wimmer, Vice President of the Parliamentary Assembly of the OSCE (Organization for Security and Co-operation in Europe) from 1994-2000, blames the US for trying to strong-arm European NATO member states to accept the geopolitical interests of the US that went far beyond the defensive agenda of NATO [Wimmer 2022].

Today, various organisations are entertained through military and secret services of the US, NATO, UK and EU member states to officially counter so-called Russian disinformation, thus building up new hostile images instead of seeking for peaceful conflict transformation [Berger 2021]. A new cold war has been started not only against Russia, but also against China and other nations that have resisted subjugation to Western claims of a so-called "rules-based order" instead of existing international laws that are in force for all nations.

That new cold war is bolstered by military measures if and when interference in internal affairs for regime change and sanctions do not play out. In the nuclear field, any quantitative disarmament was compensated by qualitative rearmament. The US continued its pursuit of nuclear primacy. According to a computer simulation published in 2006, the US was in a position to hit 99 % of the Russian nuclear missiles in a surprise attack and to intercept the remaining 1 % by the US missile defense shield (components of which are now deployed in Romania and in the Black Sea, but in the making also in Poland). This conjured up the delusion of being able to wage a nuclear war while keeping US territory sacrosanct [Lieber and Press 2006; Senghaas 2007; Ritz 2008].

This was exacerbated by the gap created by not prolonged arms control agreements, in particular the ABM Treaty that was terminated by the exit of the US. Under these conditions, the arms race has proliferated into a huge variety of "disrupting" technologies [Acton 2021]. Hypersonic gliders and nuclear-powered delivery systems are in the queue if not already ready to be deployed, and the weapon systems are increasingly being turned into so-called "autonomous systems" (e.g., for drone strikes). An escalation with nuclear weapons involving algorithms that parallel those in high-frequency trading at the stock exchange could, in milliseconds, lead to a flash crash. That would cross the tipping point and lead to perdition. Such a tipping point can clearly be avoided by reversing technology development and harnessing that technological potential for building peace.

Any of the techno-social threats mentioned here can be mitigated or resolved by negotiations between the affected parties. From a Western perspective, this would require accepting that the world of international relations has become transnational. Accordingly, unipolar relations are no longer justifiable because the existence of computerised weapons of mass destruction has the potential to blow humanity into oblivion. This was Gorbachev's point of reference. It is rational to listen to the arguments of pejoratively called "Russia understanders" and to practice diplomatic empathy. But Secondness is only a first step towards understanding the whole picture, towards accepting the absolute interests ("red lines") of any side, towards reflecting on the Third at the level of humanity. This implies a readiness of all sides to implement a win-win-situation. Any security architecture can be realised only as a common architecture, inclusive of all sides.

4.1.4.2 *ICTs for disruption or sustainability*

As of October 2021, the consensus on anthropogenic global warming and its impacts on weather patterns has exceeded 99 % in the peer-reviewed scientific literature [Lynas et al. 2021].

Scientists started early to discover climate change. The "greenhouse effect" describing warming due to the concentration of atmospheric carbon dioxide was recognised more than one hundred years ago. Edward Teller is known because of his involvement in the Manhattan Project, his opting for actual combat-use (as did also Robert Oppenheimer), as well as for his contribution to the development of fusion-based atomic weapons. He warned at a symposium on the one-hundredth birthday of the American oil industry of the American Petroleum Institute and the Columbia Graduate School of Business in 1959: "It has been calculated that a temperature rise corresponding to a 10 per cent increase in carbon dioxide will be sufficient to melt the icecap and submerge New York. [....] At present the carbon dioxide in the atmosphere has risen by 2 per cent over normal. By 1970, it will be perhaps 4 per cent, by 1980, 8 per cent, by 1990, 16 per cent if we keep on with our exponential rise in the use of purely conventional fuels" [Franta 2018].

By the 1960s, computers yielded more detailed calculations. Models began to thematise runaway positive feedbacks, concerns were voiced of a "geophysical experiment" that humanity is executing.

Today, Earth system models show sophisticated tipping elements (see subsection 3.1.2.1 point 2). Helga Nowotny, Professor Emerita of Science and Technology Studies at ETH Zurich, former President of the European Research Council, writes "that we have entered the digital Anthropocene, where non-linear connections prevail" [2021, 35].

Environmental movements garner attention and digital media formats popularise scientific results to make them intelligible to all. This has led to phrases such as "Systems Change not Climate Change" or "Extinction Rebellion". "Fridays for Future" started with the motto: "listen to science". Nonetheless, the willingness to accept those arguments cannot be presumed for all. The stance and the willingness to receive news often depends on the material living situation of the people involved. If they are educationally disadvantaged, they are prone to so-called "news deprivation" or "news avoidance". In many Western countries, the proportion of news-avoiders has risen and ranges from 15 % of media users in Denmark to 41 % in the US. Those proportions seem to correlate with the polarisation of societies, with social media use, and with low age [Reuters Institute for the Study of Journalism 2017; 2019].

The "infodemic" – as the WHO (World Health Organization) called the disinformation that went viral in the "social media" during the COVID-19 crisis – is an example for that. One point of debate is whether the SARS-CoV-2 virus is a mutation of natural origin or was manufactured by purported gain-of-function lab experiments intended to better prepare society for future outbreaks of coronavirus diseases [Wade 2021]. Importantly, however, the deep cause of the pandemic is how society relates to nature and wildlife. Most virus diseases are zoonoses, i.e. diseases that jump from vertebrate animals to humans. Vaccination campaigns represent an "end-of-the-pipe" technology but are the remedy of choice to prevent fatalities and mitigate hospitalisation as long as the deep cause remains powerful. We are being flooded with broadcasts on the severity of COVID-19, on the consequences for public health, on the shrinking capacities of hospitals, on the efficacy and safety of vaccination, which has been demonstrated several thousand million times, and on the working of different vaccines. Scientific experts provide explanations on a daily basis. Nonetheless, populist misconceptions of personal freedom and powerful economic interests torpedo solidarity with and the interests of the poor, economically challenged and less educated people. This causes governments to hesitate pursuing clear strategies and people to deny vaccination, with the result that countries are hindered from reaching the requisite immunisation level. Such a level is the implementation of requisite information the collective intelligence of societies and individuals are called upon to create in order to cope with the complexity of a pandemic. Computer models can simulate the natural dynamic of pandemics (nonlinear growth patterns) under various human interventions (restrictive physical contacts and other measures) to describe possible scenarios for future developments. Simulation experts therefore provide a scientific framework for deciding upon the societal strategy, which needs to be backed by personal responsibility. Simulations are the best means to guide social systems through times of crisis. They can be improved by more and better health data to teach societies more about the disease and the efficacy of counter-measures. Digitalisation of anonymised health data must be given priority over privacy concerns, which are often not raised when people are targeted by customised marketing.

Regarding sustainability, digitalisation has – and can have more – positive impacts for combatting the COVID crisis, for mitigating the climate crisis as well as any other ecological challenge. This is because it can promote the scientific understanding of self-enforcing dynamics that cause qualitative changes, bolster scientifically-based decision-making, and provide techno-social innovations that enable practices in harmony with the common good as an ideation and materialisation of the overall Third.

ICTs are waiting to be designed for sustainable goals.

4.1.4.3 *ICTs for misery or development*

Social development comprises cultural, political and economic development from a local to the global scale. How can ICTs support the reorientation of this evolution towards a common world society in which social relations realise justice? Numerous events can be looked upon as a hopeful foreshadowing of a GSIS.

> However, the societal development after the 1968s was not particularly conducive to the formation of strong, comprehensive, deep forces made up of agents of change in the direction of a GSIS. In the aftermath of the oil-crises in the first half of the seventies and on the eve of the eighties of the last century, the postwar boom and

> the blind trust in the steady improvement of social life conditions lost momentum. In economy, the accumulation of industry capital decoupled the increase of wages from the increase in productivity. In technology, flexible automation displaced Fordism (mass production with mass consumption). In politics, Thatcherism and Reagonomics, the destruction of the social welfare state by liberalisation, privatisation, and deregulation were introduced. In culture, the ideology of neoliberalism, of 'make your own luck', of individualism began to become hegemonic. All of that formed a pattern that connects. It was implemented by the advised response of the ruling classes to the decline of the profit rates which had accelerated because of the accumulation of capital that could not find appropriate spheres of investments. And this implementation could capitalise the weakness of the trade union and labour movements. In the nineties, the financial capital began to outweigh the industrial, 'material', 'productive' capital causing several bubble implosions. In the current crises, the transnational financial capital is targeting national economies and the politicians support it by administering austerity at the cost of the 80, 90 or even 99 per cent of the populations instead of starting a redistribution of wealth and income.
>
> Against this historical background, the development of alternative consciousness was rather improbable, since pupils were trained for working as cogs in short-sighted economic interests and were not educated for grasping the big picture. Personal competence through political education and engagement is lacking, whereas technical and business skills and (natural) science education prevails. This is the result of the economisation of education and the transformation of pupils and students in customers. [Hofkirchner 2014b, 85-86]

Bernhard Heinzlmaier, an expert in youth studies, states: "There is no longer room for education of people for critical self-reflection in institutions that are reduced to the training function and that are reduced to conveying occupational information and skills in an economically efficient way" [Heinzlmaier 2012, 5-6 – my translation]. "These institutions do not provide guidance for critical thinking nor do they provide free space for it. Bonds to society are not established" [Hofkirchner 2014b, 86].

This is all the more valid in times of Corona crisis. COVID-19 hit already distorted societies and reinforces existing iniquities. Heinzlmaier's colleague, Beate Grossegger, in addressing youth, writes: "Articulating positive utopias has become a luxury good" [Grossegger 2021 – my translation]. Heinzlmaier published a book on "Generation Corona" [2021]. The description of the situation is as follows:

> We live in the time of the uprising of the rich against the poor. Not only is the material gap between rich and poor, above and below, growing. The middle and lower classes are also increasingly marginalised in public discourse. Nobody listens to them. They are 'silenced' or arrogantly instructed and placed in the right corner, once they have mustered up all their courage and speak up.
>
> [...] Those not privileged for education who are still involved in left-wing parties or trade unions quickly notice that they are not advocating their own interests there, but only those of the functionaries and leadership cadres. They stabilise the power of political elites who only have fine, sophistic words for them and who shirk material political measures that would be in the interests of the middle and lower classes. Migration issue, minimum wage, two-class school and health system, housing and militant advocacy for bogus self-employed and precarious employees at Uber, Amazon and delivery services – none of them. Instead, there is corona and climate panic with associated chaos politics, elite support at schools and universities, tax gifts for the super-rich, trade union politics in the interests of privileged top officials, national bank employees, the labour aristocracy in the old industrial groups and, on top of that, unworldly identity politics for minorities. The vast majority of average earners and the disconnected social lower classes are left behind. [...]
>
> The young people from the middle and lower third of society are particularly frustrated politically. They distance themselves from politics. Political appeals go unheard, for example when it comes to corona measures. The middle and lower classes in particular are sceptical of the state as a whole. The state is not their state, it is the state of the economically privileged and the educated elite. Mask requirements, social distancing, event bans and the closure of shops, nightspots and discos are mostly rejected. And the willingness to vaccinate is still low. [jugendkultur – my translation]]

ICTs are said to provide possible means for social change, that is, socially progressive movements that form from the civil society and can benefit from them [Rheingold 2002; Coenen et al. 2012a; 2012b].

Initially, it was put forth that social media would be an appropriate tool for that. However, there is hardly any evidence of a movement that could ever accrue lasting success by employing social media. The problem up until the Occupy movement was that each movement was unable to become self-perpetuating because they abstained from consolidating by establishing organisational relations. This made them unable to influence politics. It was their fault: they had problems to nominate a speaker or to draft claims or to negotiate with representatives of political or economic systems.

Twitter was deemed the next appropriate tool. Like similar tools, however, it has proven adept to organise flash mobs, play cat-and-mouse games with the police or spread fake news, but not to sustain a social dynamic that truly revolutionises life.

In hindsight, ICTs have demonstrated an ability to incite regime change. They played a role in supporting "Arabellion", in "colour revolutions" such as the "Orange Revolution" in Ukraine and for other oppositional movements in post-Soviet countries, but also in the "Green Movement" in Iran and elsewhere. The point is to establish an organisation that is committed to work like an organisation. Unfortunately, those organisations that worked for regime change have been working according to a handbook that does not apply the Logic of the Third for the good. Gene Sharp is the author of that handbook, whose fourth edition was published in [2010]. A former version was used by OTPOR, the CIA-funded Serbian youth movement that brought down the Milošević government in 2000 and later became CANVAS (Center for Applied Nonviolent Action and Strategies). CANVAS is proud of having trained opposition forces in more than thirty countries. This book advocates "non-violence", a term that it does not, however, use in a pacifistic, moral or religious sense, but rather in the sense of (political) defiance. Defiance "denotes a deliberate challenge to authority by disobedience, allowing no room for submission" [Sharp 2012, 1]. Non-co-operation is an essential ingredient of defiance. According to the book, in the initial stages of the struggle, separate, limited and temporary campaigns should not focus on total non-co-operation but rather on relatively minor, non-political issues. Only after the "democratic resistance forces gained strength, [...] the goal of producing increasing political paralysis, and in the end the disintegration of the dictatorship itself" should be pursued [2010, 62]. Sharp militates against "conciliation", "compromise", and "negotiation"; against appeals "to the dictators' sense of common humanity"; against convincing them "to reduce their domination bit by bit, and [...] to give way completely to the establishment of a democracy"; against conceding to the dictators that they "may have acted from good motives in difficult circumstances", giving them "encouragements and enticements" that would allow them "to remove themselves from the difficult situation facing the country"; against offering them a "win-win" solution, "in which everyone gains something"; and against the willingness of the democratic opposition "to settle the conflict peacefully by negotiations (which may even perhaps be assisted by some skilled individuals or even another government" [2010, 9]. "When the issues at stake are fundamental", he writes, "affecting religious principles, issues of human freedom, or the whole future development of the society, negotiations do not provide a way of reaching a mutually satisfactory solution. On some basic issues there should be no compromise. Only a shift in power relations in favor of the democrats can adequately safeguard the basic issues at stake. Such a shift will occur through struggle, not negotiations" [2010, 10]. He states, to reach that aim, external, international "assistance" is allowed, including mobilising "the world public opinion [...] on humanitarian, moral, and religious grounds"; obtaining sanctions like "reduction of levels in diplomatic recognition or the breaking of diplomatic ties, banning of economic assistance and prohibition of investments"; and providing "financial and communications support [...] directly to the democratic forces" [2010, 50].

> Thus Sharp's guide to defiance is essentially the propagation of an allegedly non-violent form of regime change. With violent forms, however, non-violent forms share intransigence on the part of those who are to defy the regime. By aiming at regime change, regardless of whether it is violent or not, there is no room for granting legitimacy to the opponents. Defiance implies that those who defy represent a legitimate cause whereas those who are defied do not. Reciprocally, those who are defied regard their cause as legitimate while they do not consider the cause of those defiant to be legitimate. Therefore, changes from non-violence to violence may happen more easily than not [...]. Using defiance is not a safeguard against military war. For that reason the concept of non-violent struggle is intrinsically flawed if based upon intransigence. [Hofkirchner 2014d, 58]

Defiance and intransigence are remnants of old thinking. They are not a method of choice in bringing forward a techno-eco-social transformation of the magnitude that is required. This does not mean that rationality is ineffective. In contradistinction, a discourse is required that is conducted based on rational arguments that provide room to envision, together, a Third by which antagonisms and agonisms are elevated to the level of synergisms.

Some time ago a project proposal was formulated that applied a comprehensive perspective of designing ICTs for the whole globe. The outcome was described as "global but decentralised, democratically controlled information platform to combine online data and real-time measurements together with novel theoretical models and experimental methods to achieve a paradigm shift in our understanding of today's highly inter-dependent and complex world and make our techno-socio-economic systems more flexible, adaptive, resilient and sustainable through a participatory approach" [Bishop and Helbing 2012].

The consortium was made up of leading European researchers and other institutions and envisaged the following principal components of the platform:

> The "Planetary Nervous System" (PNS) will bring together existing and new data sources, such as demographic data, mobility and activity patterns, financial and economic data, epidemiological and other data, and enrich them creating higher-level semantic meaning. To this end it will implement a privacy-respecting and user-controlled paradigm of social data mining, featuring question-driven, self-organised, self-optimising, self-regulating, and decentralised measurements, data collection and enrichment. In summary, PNS will turn raw data into semantically meaningful information.
>
> The "Living Earth Simulator" (LES) will combine the information provided by the PNS with models to enable simulations of social, economic, technical and environmental developments, enabling a better understanding of the state of our world, and its possible futures.
>
> A "Global Participatory Platform" (GPP) will enable citizens, businesspeople, scientists and policy-makers to interact with the Planetary Nervous System, the Living Earth Simulator, and FuturICT's Exploratories. It will display the answers to their questions in new, exciting and engaging ways, using also serious games and immersive technologies. It will feature an Open World of Modelling Platform, a trusted brokerage system to bring together data producers, data consumers, and resource (e.g. computing systems) providers into a novel information ecosystem. The outcome of this work will be a deep embedding of evidence-based decision-making throughout society, enabled by large-scale adoption of the resources mediated by the GPP.

The GPP is intended "to create knowledge and collaboration opportunities that go beyond what any one user or any one team can achieve". A "Global Systems Science" is to form the theoretical basis of the platform.

> [...] "Exploratories" [...] will integrate the functionality of the PNS, LES, and GPP, and produce real-life impacts in areas such as Health, Finance, Future Cities, Smart Energy Systems, and Environment. The combination of PNS, LES and GPP will empower non-experts to interactively get answers to high-level queries about our complex techno- socio-economic system(s) on a global scale, e.g. about factors relevant for social well-being. This will lower barriers to social, political and economic participation and create an information ecosystem enabling new data- and model-based businesses. By connecting the Exploratories over time, FuturICT will eventually be able to study the interdependencies between the Environment, the Economy, our Society and Technology, and to build an integrated FuturICT platform to study global interdependencies and dynamics.

An "Innovation Accelerator" complements the platform "to speed up the development of new scientific knowledge, and achieve more rapid routes to dissemination and commercial benefit".

The promoted ICT paradigm is characterised by the following features:

- Participatory platforms as instruments to enable new forms of knowledge (co-) creation, collective awareness, and social, economic, and political participation.

- Socially interactive and socio-inspired ICT as new paradigms for self-organisation, adaptation and trustworthiness.

- A new information ecosystem and a favourable co-evolution of ICT with society.

- Privacy-respecting mining of Big Data and user control of their data as well as ethical, value-sensitive ICT (responsive and responsible).

Unfortunately, that project proposal, which was submitted to the "Future Emerging Technologies" flagship EU funding in 2012, had to compete with a proposal for modelling the human brain activity and with the nano material project on graphen, which won the qualification. It would, however, have had a much more important impact on social development than the other projects. It would have provided thoughtful deliberation on how to create meaningful technologies for making Earth a better place instead of pursuing mere economy-driven digitalisation.

From a positive perspective, none of the efforts to bring about change has been in vain. All have fostered the germ of growing political awareness. Insight into the causes of the crises has proliferated, and the discourses have come to recognise that the current crises are expressions of a progressive enclosure of all the common goods generated and utilised by actors in the whole range of social systems that make up society. Battles over reclaiming the commons can more easily be identified than before – both by the public and by social scientists. As in any evolution of self-organising systems, social evolution inheres imponderabilities, contingencies and serendipity – it is emergent, and situations might very well occur that open new windows to the future. The Third is *ante portas*.

4.1.5 *Responsible man-machine designs*

Roberto Simanowski, a German expert in German Literature and Media Studies, published two essays with MIT Press on "The death algorithm and other digital dilemmas". He then significantly expanded and revised those essays for his book in German [2020]. The death algorithm is the programme that, in case of imminent car accidents, steers the self-driving vehicle into a target of choice. Simanowski rigourously demonstrates aporias and paradoxes that cannot be solved based on programming vehicles according to utilitarian/consequentialist or deontological ethics: no programme will satisfy a universal rule acceptable for all humans. His idea is that strong artificial intelligence (AI) could provide such a solution. He believes that Deep Learning might enable strong AI not only to follow decisions reached by human intelligence but also to make its own decisions independently of human intelligence. Accordingly, strong AI might be able to make decisions that human intelligence is still unable to make because vested interests frustrate and cancel each other; strong AI might be disinterested in human particularisms and neutrally watch over humans who carry out the decisions that AI would make. Strong AI might therefore even help humanity survive. By that viewpoint, Simanowski continues the imaginary of “machines of loving grace” – a benevolent technocracy safeguarding an all-out harmony.[i]

Six years after that poem was published, Ivan Illich, who worked as a parish priest, university rector and professor in the field of Science-Technology-Society at Penn State University and commuted between Mexico, the US and Germany, created another vision – a convivial society that shapes technology for conviviality, “Tools for conviviality” as he titled his book [1973].

Both options seem to share an optimistic picture of the future development of culture and civilisation. What about the role of AI in machine designs? What about its relation with human intelligence?

The analysis of the ways of designing the technological and the societal side in suggested intelligent “man-machine” configurations

(1) reveals assimilation of both sides when the two are conflated, be it the techno-determinist reduction or the social-constructivist projection;
(2) exhibits segregation of both sides when they are disconnected, either by technocentrism or by sociocentrism or by techno/social interactivism;
(3) and concludes with an integration of both sides in techno-social systemism.

4.1.5.1 *Man-machine assimilation*

Both reduction and projection converge in propagating the following point of view: the artificial and the human are to be treated in one and the same way, that is, indiscriminately. Assimilation, however, means that either society is assimilated by technology or technology is assimilated by society (Table 4.6.a).

---

[i] In 1967, while Poet-in-Residence at the California Institute of Technology, US writer Richard Brautigan published a poetry collection. In one poem titled “All Watched Over by Machines of Loving Grace” he fancies “a cybernetic ecology | where we are free of our labors | and joined back to nature, | returned to our mammal | brothers and sisters, | and all watched over | by machines of loving grace” [quoted in Madrigal 2011].

Technodeterminism. The techno-determinist view of man-machine design can be summarised as follows: man shall be treated like machines, the human like the artificial. This view is an assimilation of the human to the artificial, entailing an assimilation of human intelligence to artificial intelligence. It encompasses the concept of the cyborg and is the motto of transhumanism. The design aims at *Homo deus* [Harari 2016] – a longing to perfect the species through artificial means, including enhancing its intelligence. As Fleissner already noted, reference to the Internet often offers the potential to realise omniscience, omnipresence, omnipotence and omnibenevolence [Fleissner and Hofkirchner 1998]. Humans shall be engineered to become optimised. "Identified 'shortcomings' of the human condition are social problems to be solved technically in contrast to the cultivation of social abilities and ambitions of humankind" [Hofkirchner 2021a, 189]. The delusion to become god-like is hubris, "but this is conjoined by Anders's humiliation, in the sense that what is human is degraded and reduced to the utilization of a narrow spectrum of technological enhancement and augmentation opportunities" [Hofkirchner 2021a, 189]. Anders claims that humans experience Promethean shame in response to things fabricated by artifice. After Copernicus, after Darwin and after Freud, Anders describes another blow dealt to humanity's sense of itself [2016].

Table 4.6.a. Conflated man-machine designs.

	man-machine designs			
conflation	**assimilation:** the human and the artificial shall be treated in-discriminately	**reduction**	**technodeterminism**	
			the human treated like the artificial	**homo deus:** hubris from humiliation (*transhumanism*)
		projection	**social constructivism**	
			the artificial treated like the human	**techno sapiens (humanoids):** humiliation from hubris

Social constructivism. While the motto of the technodeterminist approach to man-machine designs is "man shall be treated like machines", the motto of social-constructivist designs is "machines shall be treated like humans, the artificial like the human". That is the reverse assimilation of humans to machines. Humans become assimilated by adding to machines a value that is improper for machines.

> Anthropomorphism consists in attempts to perfect the artificial by making it more and more human-like, that is, humanoid robots, androids, etc., culminating in what Peppo Wagner refers to as "techno sapiens" (2016): a machine that is purportedly equivalent to the human, at least in terms of parameters taken to be essential to establish that the entity is indistinguishable from the human. This trend is not new. It has, since the beginning of informatization, been recognizable within the language used to describe features of computing, artificial intelligence and robots (consider the derivation of "computer", "robot" and the connotation of "man–machine communication" and so forth). It is also inscribed in the mass production processes that introduced fixed automation and then flexible automation, and in so-called autonomous systems. It is identifiable in debates concerning whether or not moral rules should be implemented in and for robots; though it might be argued, as philosopher Susanne Beck points out, that machines cannot rationalize on the basis of sentience, values and intuition, and cannot be taught the evaluation of infinitely contingent situations, and thus cannot act morally even if they could mimic human decisions (Stanzl 2017). This is despite attempts to devise such an algorithm [...] (see also Trappl 2015).
>
> Various related issues have arisen, for example, whether or not "acting" and "thinking" machines should be endowed with electronic "personhood" and granted rights and obligations. [...] The issue has an economic and social context. For example, whether or not children and the elderly should be treated by robots as (cheaper and available) substitutes for interaction and communication with caring human persons. Behind this issue lies the further consideration of whether androids could and should be endowed with the capacity to detect emotions in humans, and then not only simulate those emotions (as-if-emotions) but actually have them. [Hofkirchner 2021a, 190]

The claim for endowing such devices with AI to make them "autonomous" is not pure hubris. For Anders, humiliation derives from that hubris because those devices delimit the generic autonomy of humans.

4.1.5.2 *Man-machine segregation*

The segregative compartmentation of the techno/social praxiology arrives at the following recommendation: man and machine shall be treated discriminately, the artificial and the human shall be treated in discriminative ways. There are three varieties to do so: the technocentric, the anthropocentric, and the interactivistic one (Table 4.6.b).

Table 4.6.b. Disconnected man-machine designs.

	man-machine designs		
dis-connection	**segregation:** the human and the artificial shall be treated in discriminative ways	**technocentrism**	
		the artificial treated better than the human	**Übermensch ex machina:** hubristic humility (*post-humanism*)
		sociocentrism	
		the human treated better than the artificial	**pride of creation:** humble hubris
		techno/social interactivism	
		both treated on equal terms	**networks:** hubris-humility shifting

Technocentrism. The main point of technocentric design is that the machine shall be treated better than man, the artificial better than the human. Technological progress is put first. Technology is perceived as something that runs ahead while the development of society will always lag behind.

> The machine is to be perfected to be devoid of human error. If a machine is liable to failure, then it is because of errors of the operators, that is, humans, because of programming errors that are the fault of humans, or because of material defects that are, in the end, due to faults of humans, again. Machines can, in principle, and they do so in reality, outperform humans. Intelligence of machines will render the intelligence of humans obsolescent. That is the credo of posthumanism and singularitarianism – a kind of Nietzsche's Übermensch but ex machina, that is, from the machines, robots, autonomous systems, AI. [Hofkirchner 2021b, 41]

While the belief that humans are capable of constructing machines that are superhuman is hubris, such a superhuman construct would be a self-humiliation of humans. Anders talks of "arrogant self-degradation" or "hubristic humility" [Anders 2016, 50 – italics removed, W.H.].

Irrespective of whether the aspirations of post-humanists and singularitarianists can be ontologically substantiated to make them true, they are flawed praxiologically. Moreover, they would even endanger the further development of the human species because they would disrupt human autonomy. As Karl Marx had foreseen, automatisation can lead to a social state of affairs "where we are free of our labors" as Brautigan's poem of machines of loving grace insinuated. But automatisation need not lead to autonomisation. Autonomy is a term borrowed from the realm of humans and society and imposed on machines, as if a machine would be a person. Autonomy of automatons signifies, in engineers's speak, that automatons can also reach decisions. This would mean that functions of the human mind would be taken over by automatons such that the human mind would be deprived of those functions: either the automaton or the human takes the decision – only one of them can take it. Thus, Simanowski's faith in strong AI saving humankind is doomed to failure because the autonomisation of machines would cause the de-autonomisation of humans [Hofkirchner 2022].

Sociocentrism. Sociocentric design holds that the man shall be treated better than the machine, the human better than the artificial. The human shall be perfected without resorting to technology, which is treated as trumpery or even dangerous (for example AI, which might devaluate the position of human intelligence). Being the pride of creation is hubris, but a humble hubris because it tends to omit technological help.

Techno/social interactivism. As to the praxiological positions of man and machine, the third disjunction does not prioritise either side: man and machine, the human and the artificial shall be treated on equal terms. In hybrid networks, design either levels-up machines and/or levels-down humans, but the treatment of both is not reconciled on a higher level. Differences in dealing with the artificial and the human are levelled out such that none of them receives proper treatment. This is epitomised by French sociologist Bruno Latour's Actor-Network-Theory [2006].

> According to the famous saying of Latour that it is not me who shoots with the pistol but it is the pistol which (maybe better: who?) shoots with me, it is not humans who make decisions but intelligent devices whose decisions we just adapt to or execute (e.g., in the case of so-called expert systems in health care). [Hofkirchner 2021b, 41]

Such a network

> [...] meanders between hubris and humiliation. Hubris comes to the fore when the role of artefacts is equalled to (projected onto) the role of humans, while humiliation comes to light when the role of humans is equalled (reduced) to the role of artefacts. In short, we have got a hubris-humility-shimmering – not a desire at all. [Hofkirchner 2021a, 193]

4.1.5.3 *Man-machine integration*

Techno-social systemism – the view that combines technology and society – is an integrative view. Accordingly, man and machine, the human and the artificial, shall be treated according to their different properties that make them qualify for the sake of the whole (Table 4.6.c).

Table 4.6.c. Combined man-machine designs.

	man-machine designs		
	integration:	**techno-social systemic activism**	
combi-nation	the human and the artificial shall be treated according to their different properties for the sake of the whole	the human and the artificial are treated appropriately: technologies are constructed for the synergy of human actors and social systems	**the good society:** no hubris no humiliation (*alter-humanism harnessing tools for conviviality*)

The treatment is appropriate. Technologies are constructed for the synergy of human actors and social systems. Synergy means a certain constellation of organisational relations that constrains and enables the interaction of the elements to become a co-action in which every element finds its proper place [Hofkirchner 2017c, 9]. Social synergy materialises in the commons. The social organisational relations of production and provision determine how the human and the artificial shall be related. In the good society, they are related to advance the commons. "Thus, artificial devices can and should be nested in the social system" [Hofkirchner 2021a, 194]. Those are meaningful technologies.

> The default value of meaningful technology is to serve the vision of a good society, of individuals living a good life and of cultivating the common good. Such an alter-humanism instead of an old-fashioned humanism or post-humanism [...] harnesses tools for conviviality (Illich 1973). [Hofkirchner 2021b, 45]

Meaningful technologies are designed by a society that is aware of the responsibility assumed for the social functions that shall be served. There is no hubris and no humiliation. Such a design values the social Third high.

4.2 Remodelling Information Technology for the Future: the Emergence of Digital Intelligence

The analysis of the approaches to modelling the technological and the societal side yields ontological monism in the case of conflation, in particular the pairing of technomorphism as to the reduction and anthropomorphism as to the projection; it yields ontological dualism, if not pluralism, in the case of disconnection, in particular technocentrism, anthropocentrism, and techno/social indifferentism; and it yields ontological dialectics in techno-social systemism in the case of combination.

Ontology of the technological and the societal in information processes has been increasingly focussing on man-machine relationships. This is because it has been questioned by an alleged blurring of what is (to be) denominated as artificial and what is (to be) denominated as human.[j] The pivot here is the phenomenon of intelligence (Table 4.7). How is so-called "artificial intelligence" related to human intelligence [Hofkirchner 2021a; 2021b]?

> Oswald Wiener, Austrian cyberneticist, avant-gardist writer, musician, linguist and others, blames the Silicon Valley for merely simulating [Jungen 2017]. They rush into talking of "learning" and "intelligence", although all current AI machines do not go beyond "flat formalisms", he says. Those machines outclass humans in computation, but they are a "surrogate of intelligence" only. They are stupid on a high level. "The heterarchically ordered depth of human rationality cannot be realised in such a way." Humans are incapable of following a straight algorithm. The recognition process is rather a "recursive process", in which data are permanently matched with given knowledge in the background. Thinking is dependent on sensuality. [Hofkirchner 2020c][k]

---

[j] For an in-depth discussion of this question with a lot of examples see the book of the Davis Professor of Complexity at the Santa Fe Institute, Melanie Mitchell [2019].

[k] Oswald Wiener died of pneumonia in 2021 at the age of 86. The quotes are taken from a talk he gave and were published in a newspaper article. The translation my own. Wiener is famous for the novel titled "Die Verbesserung von Mitteleuropa" (The betterment of central Europe), in which he anticipated a simulation of virtual reality that was visited by

This is a criticism that techno-social systemism supports (Table 4.7).

Table 4.7. The consideration of non-being and being in techno-social systemist ontology. The (human-centred) artificial intelligence and the human/global-netizenship pro-active intelligence.

<table>
<tr><th colspan="2"></th><th>non-being</th><th>being</th></tr>
<tr><td rowspan="2">con-
flation:
monist
fallacy</td><td>reduction:
techno-
morphic
fallacy</td><td>the artificial intelligent:
sufficient condition for the human intelligent</td><td>the human intelligent:
resultant of the artificial intelligent</td></tr>
<tr><td>projection:
anthropo-
morphic
fallacy</td><td>the artificial intelligent:
resultant of the human intelligent</td><td>the human intelligent:
sufficient condition for the artificial intelligent</td></tr>
<tr><td colspan="2" rowspan="2">disconnection:
techno/social dualist (pluralist) fallacy</td><td>artificial intelligent</td><td>human intelligent</td></tr>
<tr><td colspan="2">independent existents</td></tr>
<tr><td colspan="2">combination:
techno-social dialectical ontology for a Great Transformation</td><td>the human-centred artificial intelligent:
necessary condition for a global-netizenship pro-active intelligent</td><td>the global-netizenship pro-active intelligent:
an emergent from the human-centred artificial intelligent</td></tr>
</table>

Techno-social systemism is anti-reductionist: modelling human intelligence according to artificial intelligence as a sufficient condition is fallacious.

It is anti-projectionist: modelling artificial intelligence according to human intelligence as a sufficient condition is fallacious too.

It is anti-disconnectionist: modelling artificial and human intelligence as incommensurable existents is merely another fallacy.

Techno-social systemism models AI – that is human-centred – as a condition for the emergence of a pro-active future intelligence of global netizens.

Those arguments are discussed in detail below.

---

real persons and that simulated that they would leave the simulation while they were still captured by it.

4.2.1 *Man-machine monism*

Following the conflation of the views of the praxiology of technology and society, the conflation of the views of the ontology of artificial and human intelligence is: man and machine, human and artificial intelligence, are identical inasmuch as they share the same degree of complexity. Nonetheless, the reductionist and the projectionist view differ (Table 4.8.a).

Table 4.8.a. Conflated man-machine models.

	man-machine models		
conflation	**monism:** the human and the artificial are identical, inasmuch as they share the same degree of complexity	**reduction**	**technomorphism:** the human is as complex as the artificial
		projection	**sociomorphism:** the artificial is as complex as the human

4.2.1.1 *Technomorphism*

The reductionist variety is technomorphic monism. This is because, if technodeterministic design stipulates that man shall be treated like machines, man must also be modelled as machines. Thus, man and machines share essentially the same degree of complexity: human intelligence is as complex as artificial intelligence, both are mechanisms – hence monism.

In fact, mechanisms are not complex. Mechanisms require shrinking the space of possibilities of self-organising systems such that all possibilities but one are cancelled out. Only a single option remains to be taken. "Technology as constructed by humans shall always yield a determinate output given a determinate input. Therefore, mechanisms are built that restrict the possibility space to one possibility only" [Hofkirchner 2021a, 195]. Humans can, of course, function like mechanisms. They can socially be entrained to rhythms, but their common activities can go far beyond. Their individual faculty of thinking exhibits the ability to compute and draw formal-logical conclusions, but thinking is more than that. Their body includes known biological dynamics that are based on feedback cycles. But those biotic mechanisms are present only in subordinate positions.

The technomorphic fallacy involves a concatenation of the following reduction steps:

(1) **Merism.** The essential organisational features of a society are reduced to those of individual actors – a fallacy of reduction of complexity from the social system to mere social elements of the system. That is, the collective intelligence of society is reduced to individual intelligence.

(2) **Biologism.** The essential organisational features of individual actors are reduced to those of the human body – a fallacy of the reduction of complexity from social features of the individuals to mere biotic features. That is, individual intelligence is reduced to neurological features of the brain.

(3) **Physicalism.** The essential organisational features of the human body are reduced to those of its physical substrate – a fallacy of the reduction of complexity from the biotic features of the body to mere physical features. That is, a brain feature is reduced to the physicality of neurons' working.

(4) **Mechanicism.** The essential organisational features of the physical substrate are reduced to those of mechanisms – a fallacy of the reduction of complexity from physical features that still exhibit physical self-organising capacities to mere mechanistic features that are devoid of self-organising capacities. That is, the physical functioning of neurons is reduced to mechanical functioning.

Regarding intelligence, human intelligence boils down to a mere mechanist capacity that artefacts can be made capable of.

4.2.1.2 *Sociopomorphism*

The sociomorphic ontology of man-machine models bolsters the social-constructivist approach. Since machines shall be treated like humans, they demand human-like models. Machines and humans are conflated in that any mechanism is imagined as being as complex as socially evolving humans are: machines and humans are essentially the same, hence monism.

Features such as exhibiting intelligence are projected from humans and their systems onto machines. Similar to pan-idealism or pan-psychism, the world is conceived of as a computer, albeit not of a Turing-type. This is the case in info-computationalism [Dodig-Crnkovic 2014], according to which nature is always and everywhere engaged in informational processing, called "natural computing". This implies that the term mechanism is congruent with the term self-organisation. Mechanism is animated.

The sociomorphic fallacy runs through a concatenation of the following projection steps:

(1) **Structuralism.** The essential organisational features of a society as bearer of social relations are projected onto those of individual actors – a fallacy of the projection of complexity from the social system onto mere social elements of the system. That is, collective intelligence of society is projected onto individual intelligence.

(2) **Anthropism.** The essential organisational features of individual actors as social beings are projected onto those of the human body – a fallacy of the projection of complexity from social features of the individuals to mere biotic features. That is, individual intelligence is projected onto neurological features of the brain.

(3) **Psychism.** The essential organisational features of the human body are projected onto those of its physical substrate – a fallacy of the projection of complexity from the biotic features of the body onto mere physical features. That is, a brain feature is projected onto the physicality of neurons' working.

(4) **Animism.** The essential organisational features of the physical substrate are projected onto those of mechanisms – a fallacy of the projection of complexity from physical features that still exhibit physical self-organising capacities onto mere mechanistic features that are devoid of self-organising capacities. That is, the physical functioning of neurons is projected onto mechanical functioning.

The social phenomenon of intelligence is hypostatised as being characteristic of technological artefacts.

4.2.2 *Man-machine dualism (pluralism)*

The ontological disconnection abides also by its praxiological counterpart. While the segregation postulated that the human and the artificial shall be treated discriminately, the man-machine relation is modelled on the assumption that the human and the artificial, human and artificial intelligence, are different because they are genuine entities of incomparable complexity – hence dualism (Table 4.8.b).

Table 4.8.b. Disconnected man-machine models.

	man-machine models	
dis-connection	**dualism (pluralism):** the human and the artificial are different because they are genuine entities of incomparable complexity	**technosingularism:** the artificial can become more highly complex than the human
		sociosingularism: the human is of unparalleled complexity
		techno/social indifferentism: the human and the artificial are of different degrees of complexity but interact as if independent of those differences

4.2.2.1 *Technosingularism*

Technosingular models are the purest case of anti-humanism. According to such models, machines are, in brief, superior to man. The artificial is deemed to become more highly complex than the human, and the same might pertain to AI with regard to human intelligence.

Technological development has proven that certain mechanisms can outperform human abilities in certain parameters, and these are taken as an empirical basis for the generalisation that AI would be able to outperform human intelligence and thus render the human race obsolescent. AI would be able to take decisions itself and behave autonomously.

This has not yet become true and will never come true because of ontological obstacles that are so fundamental that they can never create change.

> A decision is a judgement – the result of deliberating on grounds and these grounds do not prejudice the judgement. The judgement is an act based upon the grounds as necessary presuppositions, but it is not logically derivable like a conclusion from premises. There is more to a judgement. It is an emergent act. Emergence means that the emergent has another – a new – quality compared to that from which it emerges. No machine, no automaton, no so-called autonomous technical system, can produce emergence since all underlying information processes are, ultimately, deterministic mechanisms. So, what engineers call decision in the case of a machine is, actually, a product of mechanical determinacy devoid of deliberation of a self. Moreover, given the complicatedness of modern machines, the products of their processing have become unpredictable – therefore they were mistaken as emergent by engineers – and even, practically, not explainable (retrodictable), though every step of the algorithmic processes follows a determined rule. [...] The so-called decisions of strong AI would be random and inapt. [...] the rule of AI would devolve into a rule of technocratic dictatorship in which one arbitrary situation would be superseded by another arbitrary one [...]. [Hofkirchner 2022]

AI is itself not a self-organised system; it has no self, it is hetero-organised, externally organised. It is a patient and not an agent, as philosopher of information Rafael Capurro explained [Capurro 2012]. "Agency is a property of self-organizing systems for which less-than-strict determinism holds, but not of mechanisms for which strict determinism holds" [Hofkirchner 2021a, 198]. The IEEE Global Initiative on Ethics of Autonomous and Intelligent Systems (A/IS) published a comprehensive document on ethically aligned design. It states, in Chapter Classical Ethics in A/IS, with reference to Capurro [2012] and Hofkirchner [2011b]:

> Of particular concern when understanding the relationship between human beings and A/IS is the uncritically applied anthropomorphic approach toward A/IS that many industry and policymakers are using today. This approach erroneously blurs the distinction between moral agents and moral patients, i.e., subjects, otherwise understood as a distinction between 'natural' self-organizing systems and artificial, non-self-organizing devices. As noted above, A/IS cannot, by definition, become autonomous in the sense that humans or living beings are autonomous. With that said, autonomy in machines, when critically defined, designates how machines act and operate independently in certain contexts through a consideration of implemented order generated by laws and rules. In this sense, A/IS can, by definition, qualify as autonomous, especially in the case of genetic algorithms and evolutionary strategies. However, attempts to implant true morality and emotions, and thus accountability, i.e., autonomy, into A/IS blurs the distinction between agents and patients and may encourage anthropomorphic expectations of machines by human beings when designing and interacting with A/IS. [The IEEE Global Initiative 2019, 41]

That ontological distinction that is accessible through theorising cannot be explained away by creating the false empirical appearance of intelligence based on observing behaviour. The Turing test, for instance, proves the opposite, namely how easily human comprehension can be fooled.

The point at which proponents expect AI to leave human intelligence behind is called the singularity and it would be reached by technology itself. Since technology does not exhibit a self, it will not be in the position to propel the emergence of a meta-level as a Third. There is no Second and there are no Firsts to trigger quality leaps as foreseen by the course of social evolution through the agens of human actors.

4.2.2.2 *Sociosingularism*

The sociosingular approach to man-machine models articulates, in brief, that man is superior over and above any machine. Man is not a machine. "The human is deemed the pride of creation and is not in need of some other worldly goal beyond improvements by purely social (cultural, 'spiritual') improvisations" [Hofkirchner 2021a, 197]. Human intelligence is an entity with unparalleled complexity.

It goes without saying that human agency can manifest qualities that computers cannot: "purpose, objectives, goals, telos, caring, intuition, imagination, humour, emotions, passion, desires, pleasure, aesthetics, joy, curiosity, values, morality, experience, wisdom and judgement" as Adrian Braga and Robert K. Logan argue [2021, 137]. All these features make up a sense of self that has developed during millions of years of natural and, finally, social evolution. It would, however, be erroneous to disclaim that AI can extend human intelligence merely because AI is not intelligent itself. "Stripped of these qualities as is the case with AI all that is left of intelligence is logic" [Braga and Logan 2017]. AI can simulate the logical features of human intelligence and can thus be applied to improve (this part of) human intelligence.

4.2.2.3 *Techno/social indifferentism*

The third variety of disconnected man-machine models might be labelled pluralistic. A plurality of entities – human and non-human intelligences of unknown complexities – assemble to interact as if their differences in complexity would not matter. Thus, the term pluralism is used here to describe the situation of juxtaposed differences, of differences that do not make a difference.

Latour's ANT network is conceived of as a flat ontology that obscures the effective function of the interaction in order not to hypostatise the uniqueness of either man or machine. Both are modelled as "actants", participating equally in a network, in which each is as causative as any other. "Consequently, the capability of social actors to control technology when producing or using it is conflated with the affordance of artificial devices" [Hofkirchner 2021a, 198].

Sociomaterialism seems to go no further when conceiving generic "intra-action" of agents with their ecologies, in which primacy is missing for discursive practices or materialised phenomena [Barad 2012; Suchman 2007].

4.2.3 ***Man-machine dialectic***

The dialectic combination of techno-social systemism underlines the evolutionary point of view when conceiving of man and machine, human and artificial intelligence, and their configuration. They differ in complexity and constitute a united complex with asymmetrical roles (Table 4.8.c).

Table 4.8.c. Combined man-machine models.

	man-machine models	
combi-nation	**dialectics:** the human and the artificial differ in complexity and constitute a united complex with asymmetrical roles	**techno-social systemic evolutionism:** techno-social systems are emergent from society whenever technological mechanisms are inserted in social systems

Techno-social systems emerge from society whenever technological mechanisms are inserted in social systems. Digital intelligence emerges from human intelligence whenever AI supports it.

The point of departure is the following:

> Since information generation is a process that allows novelty to emerge, it is worth noting that information generation is not a mechanical process and thus defies being formalised, expressed by a mathematical function, or carried out by a computer. It is only in the case of a mechanical process, that methods of mental transformation apply so as to unequivocally lead from a model of the cause to the model of the effect. These intellectual methods are provided by formal sciences like formal logic, mathematics, or computer science; they involve the deduction of a conclusion from its premises or the calculation of a result or a computer operation (Krämer 1988). Mechanical processes can be mapped onto algorithmic procedures that employ clear-cut and unambiguous instructions capable of carrying out by the help of computers as universal machines. But the generation of information escapes algorithmisation, in principle.
>
> [...] [Algorithmic information theory] does not cover the whole range of what the phenomenon of information embraces. In particular, it must fail to reflect novelty as essential quality of information. Deductions, by definition, don't yield novelties, algorithms, by definition, can't do it either, nor can computation, by definition, do it. [Hofkirchner 2011b, 193]

Having said that, AI is clearly subject to the above restrictions, whereas human intelligence is not. Both are products of evolution in that they are identical but also different. "Ontologically, humans and society are the product of physical, biotic and social evolution; the machine is a product of humans and society" [Hofkirchner 2020c, 1].

Physical difference. The human and the artificial differ in the physical aspect:

> Humans and society, on the one hand, and machines, on the other, share the fact that they are entities, and embrace processes that belong to the physical realm. However, they differ essentially with regards to the specifics of their being physical and behaving physically. After Rafael Capurro, humans and society are an *agens*, whereas a machine is a *patiens* [...], which is indicated by the following:
>
> • Humans and society are able to organise themselves, that is, to build up order by using free energy and dissipating used-up energy, whereas machines cannot self-organise;

> • Humans and society are made up of elements that produce organisational relations that constrain and enable synergy effects and they can constitute superordinate systemic entities, whereas machines are made up of modules that are connected in a mechanical way;
>
> • Humans and society function on the basis of less-than-strict determinacy, which yields emergence and contingency, whereas machines are strictly deterministic and cannot behave in an emergent or contingent manner. [Hofkirchner 2020c, 2]

Biotic difference. The human and the artificial differ in the biotic aspect:

> Humans and society are physical entities and activate processes that belong to the biotic realm. Machines may, but do not need to, have parts that belong to the biotic realm. Even in cases where they do so, they differ essentially in quality. Humans and society are autonomous agents [...], whereas machines are heteronomous mechanisms that cannot show any degree of autonomy, as follows:
>
> • As with any living system, humans and society are able to maintain their organisational relations by the active provision of free energy, whereas machines cannot maintain themselves;
>
> • As any living system, humans and society are able to make choices according to their embodiment, their embedding in a natural environment and the network of conspecifics, whereas machines cannot choose;
>
> • As any living system, humans and society are able to control other systems by catching up with the complexity of the challenges they are faced with by the other systems, whereas machines cannot catch up with complexity and are under control by organisms. [Hofkirchner 2020c, 2]

Social difference. The human and the artificial differ, finally, in the social aspect:

> Humans and society are not only physical and biotic, they are the only physical and biotic systems on Earth that belong to a specific, the social realm, too. They are, essentially, social agents, that is, actors. Machines are social products, artefacts, that are made by actors, but they do not possess the agency of actors. This is implied by the following:
>
> • Humans in society constitute – by action, interaction and co-action with other actors – social agency that reproduces and transforms the structure of the social system (social relations), that, in turn, enables and constrains the social agency,

> whereas machines do not partake in the constitution of society but support the action, interaction and co-action of actors.
>
> •Humans in society provide the commons as effects of social synergy, whereas machines support the provision of commons and pertain themselves to the commons.
>
> • Humans in society are the driving force of social evolution, including the evolution of culture, polity, economy, ecology and technology, whereas machines are driven by social evolution. However, they can even play a supportive role in changing the quality of the social system.
>
> • Humans in society reflect upon the social structure, whereas machines do not deliberate but support the thought functions of actors.
>
> • Humans in society set off the transition into actuality of a societal option of choice out of the field of possibilities, whereas machines do not directly trigger emergence. [Hofkirchner 2021b, 44]

Complementation of human and artificial. Note here that the human and the artificial join together according to their properties – according to the values, norms and interests regarding the social functions of the beneficiaries, and to the affordances regarding the technical functions of the tools in support of the social functions. Of course, these roles are asymmetrical.

> The mechanism itself works by strictly deterministic means to achieve the goals set by the social system actors. The social system as a whole boosts its self-organisation. This alternative integrates the machine with the human such that the digitised social system stays in command, and AI serves as a tool for humane purposes. [Hofkirchner 2020c, 3]

The role of AI in the techno-social system context is to mediate collective intelligence of society. Digital intelligence can then become the label of the societal intelligence when undergirded by AI. Importantly, the social factor is included in every techno-social system. Technology alone is an abstraction and does not even represent a system because a system in the new paradigm is defined by manifesting a new quality vis-á-vis its elements, thereby requiring self-organisation and emergence. The mechanism introduced into the social system does not change its mechanistic working. But it does change how the social function proceeds to a degree that new outcomes can be achieved. The Third that is changing is the social system Third, which is mediated by the technical means. "What is labelled AI, is nothing that can become independent and achieve a life of its own. However, it promotes the intelligence of the social system" [Hofkirchner 2021b, 44-45].

That context helps grasp why a singularity of a single supra-human AI is deemed impossible. According to complex systems expert Francis Heylighen [2015; 2016], intelligence will be distributed via societal actors. Cyber technology merely connects them. The emergence of a global brain remains rooted in humans. "From this dialectical point of view, what is in statu nascendi is a social suprasystem that would be global, notwithstanding the technological infrastructure of a global brain" [Hofkirchner 2021b, 45]. Digital intelligence is the type of intelligence needed to create that social and other information that is required to envision a Global Transformation. That intelligence takes advantage of digitalisation, a technological trend that can perfectly suit the needs of the Anthropocene.

Helga Nowotny is certain that a lesson can be learned from the COVID crisis. That lesson is a better understanding tipping points that, when crossed, prove detrimental to social evolution. This understanding was conveyed to the public by trust in the digital machines employed by experts to run simulation models of different developments.

> Just as the birds react in their singing to the synthetic voices they begin to recognize as birds to wing with, so humans react to predictive algorithms that tell them what to expect in the future. The purpose of many algorithms is to make predictions aimed at prevention. They are designed to channel social behaviour in a direction with presumed benefits for the individual and society. [...] The idea of prevention implies a normative purpose, as prevention aims to avoid harm. In contrast to the precautionary principle, which delays or forestalls action, prevention demands action to avoid harmful consequences in advance [...] [Nowotny 2021, 159-160]

That is true. "Prevention has become a normative project" [160], Nowotny states, even if the trustworthiness of predictive algorithms needs to be guaranteed by regulating a handful of monopolies, which means a domestication of those algorithms. Such algorithms "have to be hauled in from the unregulated wilderness in which they are free to roam and to destroy as they, or rather those who own them, please" [161].[1]

The prevention of dangerous tipping points is compatible with the self-limitation endorsed by Illich [1973] and recommended for the mindset to be adopted by global citizens. Nevertheless, prevention is, in essence, counter-active. The techno-social system approach is, in principle, pro-active. This is because there are also tipping points to be included that mark the reinforcement of positive developments. Such developments need to be furthered for the good, and must be explored. An example is increasing vaccination to decrease contagion. This would lead to lasting collective immunisation before every individual got vaccinated – a system effect. This COVID example can be extended to the passing of quorums that would shift the balance in favour of the social demands of progressive movements for science and technology, environmental as well as economic, political and cultural transformations.

Global citizens, connected via digital means, are world netizens and their collective intelligence can become the digital intelligence that the new world society urgently needs.

4.3 Reframing Information Technology for the Future: The Emergence of Digital Ingenuity

The first step and the second step above were to conclude the praxiological and ontological discussion of approaching IT for intelligence in the

[1] Nowotny touches a crucial point. The problem with the power of major firms and transnational corporations is that capitalism allows them to prioritise their private profit interests and to put aside the common good. Politics must gain ground and enforce the general interest against them. This holds true not only for the Silicon Valley Big Five but also for, e.g., the pharma industry, which has lost reputation during the COVID crisis. The mistake of conspiracy theories is that they overlook the fact that, despite profit interests, the exchange value of products and services does not forfeit their use value. Despite all correct and justified criticism of the pharma industry, the rejection of vaccination is not the solution. Techno-social systemism elucidates the dialectic relationship between technology and society, which is not subject to strict determinism.

Anthropocene. The remaining step is to discuss the epistemological aspects.

Table 4.9. The consideration of the apparent and the essential in techno-social systemist epistemology. The (assessment-vested) engineerable and the social-and-human-scientific/global-visioneering ingenious.

		apparent	essential
con-flation: cross-disci-plinary fallacy	**reduction: techno-universalist fallacy**	**the engineerable:** sufficient condition for the social-and-human-scientific ingenious	**the social-and-human-scientific ingenious:** resultant of the engineerable
	projection: social universalist fallacy	**the engineerable:** resultant of the social-and-human-scientific ingenious	**the social-and-human-scientific ingenious:** sufficient condition for the engineerable
disconnection: techno/social pluri-disciplinary fallacy		**engineerable**	**social-and-human-scientific ingenious**
		incommensurable knowledge	
combination: techno-social trans-disciplinary epistem-ology for a Great Transformation		**the assessment-vested engineerable:** necessary condition for a global-visioneering ingenious	**the global-visioneering ingenious:** an emergent from the assessment-vested engineerable

Praxiology is based upon ontology. Accordingly, praxiology necessitates an ontology that suits it and, in turn, ontology needs to be based upon an epistemology that suits it. Frameworks of the engineering sciences are apparent, and social and human science frameworks are also essential. The conflation of both suffers from what is called here cross-disciplinarity, including the reduction to techno-sciences and the projection of the social and human sciences, both expressing a fallacious universalist approach. The disconnection is made up of relativist mono- or mixed disciplinary methodologies and the combination of the apparent and the essential accounts for transdisciplinarity (Table 4.9).

As always, the apparent and the essential are each seen by the fallacies as either a sufficient condition for the other or as no condition at all. The engineerable results in the social-and-human-scientific ingenious, or vice versa, or none of them depends on the other. Techno-social systemist epistemology is what prepares the engineering as a necessary condition for the emergence of the social and human science ingenious. It does this by adding an assessment and design cycle to the first and global visioneering to the latter.

This warrants a more detailed discussion.

4.3.1 *Man-machine cross-disciplinarity*

In agreement with the monist postulate that man and machine are ontologically identical (because they share the same degree of complexity), they are investigated here by identical methods. Identical means that methods of one discipline that apply to one side are deemed able to cross methods of the other discipline, which also apply to the other side. Such a crossing transgresses the boundaries of the own discipline and extends its scope – hence cross-disciplinarity. This can be accomplished in two ways (Table 4.10.a).

Table 4.10.a. Conflated man-machine frames.

	man-machine frames		
	cross-disciplinarity: social and human science and engineering need identical methods	**reduction**	**techno-universalism:** engineering methods are sufficient for social and human science
conflation		**projection**	**social universalism:** social and human science methods are sufficient for engineering

4.3.1.1 *Techno-universalism*

The first variety of such cross-disciplinarity is the reductionist one. It universalises the scope of engineering methods to allegedly suffice social and human investigation. The conception of the artificial is assumed sufficient for comprehending social forms. "Methodologies that are usually built for technological research cover social phenomena. Thus, they convey expectations of technicality when applied in inquiries into social phenomena" [Hofkirchner 2021b, 36].

> In order to combine social science with engineering science, representatives of the latter might be inclined to reduce that which is human to that which is engineerable: man is deemed a machine. Operation Research, Cybernetics, Robotics, Mechatronics, the fields of Artificial Intelligence and so-called Autonomous Systems, among others, are liable to cut the understanding of man who is a social being free from the understanding of social relations; the conception of the human body free from the conception of individual actors; and conceiving of mechanics free from conceiving of the organism. Mechanical architectures and functioning that are constituents among others of human life structures and processes are analysed and hypostatised as sufficient for the comprehension of man [...]. [Hofkirchner 2017c, 6].

> The methodology in question looks for mere mechanical ties in social phenomena. That is, if you start with a frame made for mechanical phenomena and cut across social phenomena without accepting the role of non-mechanical connections, you will end up in the model with mechanical conceptualizations only. [Hofkirchner 2021a, 202]

Social and human science is reduced to engineering. Human intelligence is researched as if an artefact.

4.3.1.2 *Socio-universalism*

The second cross-disciplinary variety is the projective universalisation of the scope of social and human science methods to allegedly suffice engineering science.

> [...] representatives of social sciences – not unlike those of other disciplines – might share a predilection to understand the whole world, including artifactual mechanics, by projecting characteristics of the social world onto the former: the machine is deemed man-like. Actor-Network Theory, Deep Ecology, New Materialism, Info-Computationalism and others are prone to blur the boundaries between the social sciences and the sciences of living things, between the latter and the sciences of physical things and, eventually, between the sciences of physical things and engineering sciences by attributing human features to any of those non-social disciplines. The conception of social forms is thought necessary for the comprehension of everything. [Hofkirchner 2017c, 6]

When speaking of artificial phenomena resembling human intelligence, social universalism researches it as if it were human intelligence. “That is, it looks upon technical phenomena as if they were social ones and in doing so it carries over to them expectations that they would show what social phenomena are showing" [Hofkirchner 2021b, 36]. Social and human science is projected onto engineering.

4.3.2 *Man-machine pluridisciplinarity*

The dualistic and pluralistic ontology is reflected by a pluridisciplinary epistemology consisting of two kinds of monodisciplinarity and multi- as well as interdisciplinarity. The credo is: social and human science and engineering each require methods of their own. Again, three possibilities conform to that credo, two particularistic ones and one that is relativistic (Table 4.10.b).

Table 4.10.b. Disconnected man-machine frames.

<table>
<tr><th></th><th colspan="3">man-machine frames</th></tr>
<tr><td rowspan="3">dis-connection</td><td rowspan="3">pluri-disciplinarity: social and human science and engineering need methods of their own</td><td rowspan="2">two kinds of mono-disciplin-arity</td><td>techno-particularism: engineering needs pure engineering methods</td></tr>
<tr><td>socio-particularism: social and human science needs pure social and human science methods</td></tr>
<tr><td>multi-, inter-disciplin-arity</td><td>techno/social relativism: social and human science and engineering need a mix of particular methods</td></tr>
</table>

4.3.2.1 *Technoparticularism*

Technoparticular frames are reflective of technosingular models of machines being superior to humans. They assert that engineering requires pure engineering methods. The artificial is construed as distinctive over and against social forms. It is monodisciplinary. Monodisciplinarity means intradisciplinary research, it delves within one and the same discipline.

> Mechanical phenomena require only mechanistic frames. No attention is given to social issues. The focus allows the concentration on artefacts that might be higher complex than current man/society. [Hofkirchner 2021a, 203]

Such a methodological choice is made at departments of computer science and their applications throughout the world. "It is nourished by the condition of competitive excellence in one's own discipline and AI is one of the important fields and it has been diversifying into related fields like Autonomous Systems, Deep Learning etc." [Hofkirchner 2021b, 39].

4.3.2.2 *Socioparticularism*

Socioparticular frames reflecting sociosingular models of man being superior to machines adopt the position that social and human science requires pure social and human science methods. It "gives technological issues no attention. Thus, it does not care about artificial intelligence" [Hofkirchner 2021b, 39]. Social forms are construed as distinctive over and against the artificial.

4.3.2.3 *Techno/social relativism*

Techno/social relativist frames are the third permutation of disciplinary independence. They mirror the juxtaposition of man and machine in techno/social indifferentism in that they promote the juxtaposition of social and human science (methods) and engineering (methods). They co-exist, each as valuable as the other, without there being grounds for attributing supremacy in terms of what would count as social forms and what as technological forms. Both social and artificial forms are construed side by side.

Multidisciplinarity.

> *Multidisciplinary* [...] *research* includes several separate disciplines, e.g., when researchers from different disciplines work together on a common problem, but from their own disciplinary perspectives. [Burgin and Hofkirchner 2017, 2].

Thus, multidisciplinarity indicates a rather undeveloped state of scientific collaboration.

Interdisciplinarity.

> *Interdisciplinary research* involves interaction and coordination between several disciplines aimed at the development of knowledge in these disciplines, e.g., when researchers collaborate transferring knowledge from one discipline to another and/or transforming knowledge of one discipline under the influence of another discipline. [Burgin and Hofkirchner 2017, 3].

Despite cursory exchanges at points of intersection, however, disciplines that practice interdisciplinarity maintain reciprocal exclusivity, without significant change.

The arbitrariness and relativity of the encounter of human intelligence and AI will not transform into a consistent approach.

4.3.3 *Man-machine transdisciplinarity*

The epistemological approach of techno-social systemism resonates with the evolutionism of the combined man-machine model that constitutes a united ontological complex. The call is to interpret social and human science and engineering as systemic complements of a single methodology that comprises both disciplines from a meta-level perspective. The understandings of social and artificial forms must be melded into the understanding of the bigger picture of the techno-social system (Table 4.10.c).

Table 4.10.c. Combined man-machine frames.

	man-machine frames	
combination	**trans-disciplinarity:** social and human science and engineering need a single methodology, comprising both on a meta-level	**techno-social systemic pro-active ingeniousness:** social and human science and engineering are put together by a cycle of technology assessment and social systems design in order to promote digital ingenuity of global netizens

Such a framework can successfully be installed by turning to transdisciplinarity.

> *Transdiciplinary research* encompasses problems from different disciplines but goes on a higher level than each of these discipline goes. In other words, transdisciplinarity treats problems that are at once between the disciplines, across the different disciplines, and beyond any of the individual disciplines involved. It is aimed at understanding of broad spheres of the world directed at the unity of knowledge. [Burgin and Hofkirchner 2017, 3]

This means

> [...] an approach that assumes an interrelation of both disciplines in a systemic framework that grants (relative) autonomy to each of them according to their place in the overall framework. Both disciplines complement each other for the sake of a greater whole. That greater whole is achieved by framing both disciplines in a systems perspective, that is, by framing them as part of systems science. As such, social and engineering sciences combine for a common understanding of the systemic relationship of society and technology – of emerging techno-social systems. They make use of systems methodologies for and technology – of emerging techno-social systems. They make use of systems methodologies for empirically studying social systems and the artifactual in the context of technological applications implemented by social systems design. By doing so, they can form a never-ending cycle in which each of them has a determinate place: social systems science can inform engineering systems science by providing facts about social functions in the social system that might be supported with technological means; engineering systems science can provide technological options that fit the social functions in the envisaged techno-social system; social systems science can, in turn, investigate the social impact of the applied technological option in the techno-social system and provide facts about the working of technology. The social and the engineering parts of techno-social systems science are coupled so as to promote an integrated technology assessment and technology design cycle in a transdisciplinary sense. [Hofkirchner 2017c, 6-7].

That cycle also includes social system assessment and social system design because the techno-social system is, essentially, a specification of a social system. The performance of the technical as well as of the social functions are subject to assessment. They are assessed to induce adjustments of technology as well as of the whole social system in the next cycle.

There is one important point to recognise when understanding the combination of the human and the artificial in one common system. This is that there are no seamless connections from one to the other and back, but rather interfaces that are punctuated by processes of emergence (see analysis of the functioning of techno-social systems in subchapter 4.1.2). This is not a deficiency but rather a sufficiency to harness human creativity; this is because room is given to the mediation of variable causations. In any of those emergent processes, a meta-level option is actualised, a real Third is established that marks a qualitative leap in the transformation from the machinic, mechanist operating of the technical part to its ingenious take by the social whole.

Levels of abstraction. The specific transdisciplinary combination of social and human science and engineering depends on the levels of abstraction. Three categories of techno-social systems including IT can be identified. This categorisation makes use of the term "cyber" as recently used in "cyber-physical systems":

(1) **cyber-social systems** (e.g., in virtual communities, social networking, computer-assisted co-operative work) with actors mediated by cyber-technologies (computerised information technologies, today called digital technologies); according to the cyber-technologies they use, they can be broken down into the further subcategories of
  (a) cognitive,
  (b) communicative and
  (c) co-operative cyber-social systems;
(2) **cyber-organismic systems** (e.g., in human enhancement); and
(3) **cyber-physical systems** (e.g., in the Internet of Things) with organismic (living) and physical (material) agents and artefacts, mediated by cyber-technologies, as elements of the cyber-social systems.

Pro- and counteraction. This calls for returning to the so-called predictive algorithms that Nowotny identifies as the most significant application of AI in the digital time. Accordingly, humans need to accept that "no algorithm is able to predict the unpredictable" [2021, 106]. The reason is because mechanisms in the virtual world are no match for the dynamics of the real world. Even social sciences cannot predict social dynamics, which seems unclear to many social scientists themselves. What they can do is to think out paths that the development may take from the here and now by considering conditions advantageous and disadvantageous to one or another tendency. That is precisely what simulation scientists do when they elaborate scenarios of possible trajectories towards different future states of the social system dependent on the development of influencing factors. That some of those scenarios will not have come true in hindsight – in particular unfavourable ones – is owed to the "self-defeating prophecy": they could not come true because they were prevented (e.g., lockdowns prevented the spread of the SARS-CoV-2 virus). This was the purpose of the simulation.

At least equally valuable, however, is the task of pro-action: there are favourable scenarios too, and better trajectories can be simulated with different influencing factors (e.g., a strategy below 1.5 or below 2.0 °C Earth temperature rise might be taken). That approach helps algorithms to visioneer trajectories to real, concrete utopias. They can provide the scientific basis for a societal deliberation on which path to take. "But an AI does not know, nor can it predict, a future which remains inherently uncertain and open" [Nowotny 2021, 162]. The path selected will emerge from a process in which actors who are social by nature partake. It will not be the digital device that decides, it will be the digital ingenuity of humans that will guide humanity into the future.

An unfettered scientific understanding of human intelligence, of artificial intelligence and of their complementation to an ideational Third is what enables techno-social systemic transdisciplinarity to pave the way for tailored tools promoting the materialisation of a Great Transformation.

4.4 The Principle of Digital Humanism

In the age of a Great Transformation, critical emergentist techno-social systemism offers a theoretical concept for philosophy-of-technology based science and technology studies, technology assessment and technology design, in particular, for ICTs and Society. This concept provides cornerstones of a Critical Techno-social systems Design Theory (CTDT) for shaping the future, in particular, the digital future, thereby fleshing out the anthropocenic tenets of a Critical Information Society Theory (CIST) discussed in Chapter 3.2. It aims at harnessing the trend towards value-based, human-centred, assessment-vested IT coined "Digital Humanism". Techno-social systemism does this by updating and extending that Digital Humanism with regard to the existential risks and opportunities of humanity that have not been explicitly considered. By doing so, it adds technological imperatives that underpin the social-informational imperatives of information society. The imperatives for shaping techno-social systems provide objective conditions for empowering the subjective factor to realise the social-informational imperatives.

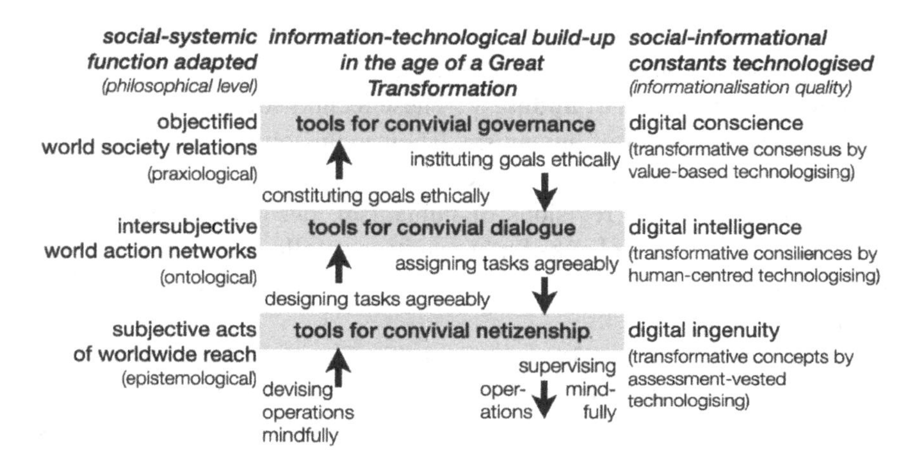

Figure 4.3. Information-technological build-up in social systems in the age of a Great Transformation.

Figure 4.3 illustrates the demanded information technology build-up in the age of a Great Transformation. That build up is an add-on to the social-informational build-up of imperatives. This Figure differs from Figure 3.23 in the designation of the levels in the middle column and the whole right column, which show the qualities of not only informatised but also informationalised, adapted, social-informational constants:

(1) **The imperative of digital conscience.** The highest level is built by tools for convivial governance that technologise the constitution and institution of consensualised transformative goals. This technologisation is value-based such that the planetary normativity is infomationalised into the techno-social entity of digital conscience.

(2) **The imperative of digital intelligence.** The next-lower level is the level of tools for convivial dialogue. The consilient designing and assigning of transformative tasks is supported by human-centred technologies so as to yield digital intelligence as a new techno-social quality of planetary discourses.

(3) **The imperative of digital ingenuity.** The bottom level encompasses tools for convivial netizenship. The technologisation of devising and supervising transformative operations conceptually integrates an assessment and design cycle for continual adjustment of digital ingenuity as planetary reflexivity.

Humanism revisited, update V. These techno-social imperatives are an integral component of humanism. Their inclusion ultimately turns the stepwise revised humanism updates I to IV into digital humanism as update V. The earlier revisions of humanism, designed to enable it to respond to the existential risks for humanity, are followed by this final step that primes techno-social innovations to exploit existential opportunities while combating the risks.

Under the title of a "Good AI Society", Floridi and co-authors dealt with four "core opportunities for promoting human dignity and human flourishing offered by AI, together with their corresponding risks" [Floridi et al. 2018, 690]:

(1) **Who we can become.** The opportunity to enable autonomous self-realisation by means of AI could be underused or over- and misused and thus potentially devalue human abilities.

(2) **What we can do.** Human agency can be enhanced with AI. But this opportunity could also be underused or over- and misused and remove human responsibility.

(3) **What we can achieve.** The possibility of AI to increase individual and societal capabilities could, again, be underused or over- and misused, thus reducing human control.

(4) **How we can interact with each other and the world.** AI can be used to cultivate societal cohesion without eroding human self-determination. However, under- or over- and misuse could lead to such an erosion.

Screening the literature on ethical implications of AI, those authors unify the principles into one framework, taking the four principles commonly used in bioethics as a starting point and adding a fifth one [Floridi at al. 2018, 695-700]:

(1) **Beneficence.** This is about promoting well-being, preserving dignity, and sustaining the planet.

(2) **Non-maleficence.** This is about privacy, security and awareness for limitations of AI.
(3) **Autonomy.** This is about "striking a balance between the decision-making power we retain for ourselves and that which we delegate to artificial agents" [Floridi et al. 2018, 698].
(4) **Justice.** This is about promoting prosperity and preserving solidarity.
(5) **Explicability.** This is about intelligibility concerning how AI works, and about accountability concerning who is responsible for the way AI works.

Clearly, the approach towards designing and developing digital artefacts that the authors share is a humanist approach that is smartly confluent with the techno-social systemism put forward here.

The term Digital Humanism was apparently first coined in a Gartner Special Report to signify "the recognition that digital business revolves around people, not technology", in order to make "employee capabilities translate into product, service and market gains" [Gartner Research 2015]. It was then used by German philosopher Julian Nida-Rümelin and Nathalie Weidenfeld in a book about ethics for the age of AI [Nida-Rümelin and Weidenfeld 2018] and subsequently taken up by then-Dean of the Faculty of Informatics at the TU Wien, Hannes Werthner, who convened a workshop in 2019. The result was The Vienna Manifesto on Digital Humanism.

> This manifesto is a call to deliberate and to act on current and future technological development. We encourage our academic communities, as well as industrial leaders, politicians, policy makers, and professional societies all around the globe, to actively participate in policy formation. Our demands are the result of an emerging process that unites scientists and practitioners across fields and topics, brought together by concerns and hopes for the future. We are aware of our joint responsibility for the current situation and the future – both as professionals and citizens. [Vienna Manifesto]

It demands the following:

> We must shape technologies in accordance with human values and needs, instead of allowing technologies to shape humans. Our task is not only to rein in the downsides of information and communication technologies, but to encourage human-centred innovation. We call for a Digital Humanism that describes, analyses, and, most importantly, influences the complex interplay of technology and humankind, for a

better society and life, fully respecting universal human rights. [Vienna Manifesto on Digital Humanism]

The manifesto proclaims eleven core principles:

Digital technologies should be designed to promote democracy and inclusion. This will require special efforts to overcome current inequalities and to use the emancipatory potential of digital technologies to make our societies more inclusive.
Privacy and freedom of speech are essential values for democracy and should be at the center of our activities. Therefore, artifacts such as social media or online platforms need to be altered to better safeguard the free expression of opinion, the dissemination of information, and the protection of privacy.
Effective regulations, rules and laws, based on a broad public discourse, must be established. They should ensure prediction accuracy, fairness and equality, accountability, and transparency of software programs and algorithms.
Regulators need to intervene with tech monopolies. It is necessary to restore market competitiveness as tech monopolies concentrate market power and stifle innovation. Governments should not leave all decisions to markets.
Decisions with consequences that have the potential to affect individual or collective human rights must continue to be made by humans. Decision makers must be responsible and accountable for their decisions. Automated decision making systems should only support human decision making, not replace it.
Scientific approaches crossing different disciplines are a prerequisite for tackling the challenges ahead. Technological disciplines such as computer science / informatics must collaborate with social sciences, humanities, and other sciences, breaking disciplinary silos.
Universities are the place where new knowledge is produced and critical thought is cultivated. Hence, they have a special responsibility and have to be aware of that.
Academic and industrial researchers must engage openly with wider society and reflect upon their approaches. This needs to be embedded in the practice of producing new knowledge and technologies, while at the same time defending the freedom of thought and science.
Practitioners everywhere ought to acknowledge their shared responsibility for the impact of information technologies. They need to understand that no technology is neutral and be sensitized to see both potential benefits and possible downsides.
A vision is needed for new educational curricula, combining knowledge from the humanities, the social sciences, and engineering studies. In the age of automated decision making and AI, creativity and attention to human aspects are crucial to the education of future engineers and technologists.
Education on computer science / informatics and its societal impact must start as early as possible. Students should learn to combine information-technology

> skills with awareness of the ethical and societal issues at stake. [Vienna Manifesto on Digital Humanism]

A workshop on questioning Transhumanism at the 2017 Summit of the International Society for the Study of Information (IS4SI) resulted in an edited volume [Hofkirchner and Kreowski 2021], which was followed by contributions at the 2019 IS4SI Summit [e.g., Hofkirchner 2020a; 2020c]. Therafter, a workshop at the 2021 IS4SI Summit addressed the missing link in Digital Humanism – the link to the global challenges [Hofkirchner and Kreowski 2022].

Global challenges teach humanity self-limitation in the sense of Illich's tools for conviviality. In this wake, Klaus Kornwachs, philosopher of technology, provides eight rules for using machines:

> 1. Never use a decision-making system that substitutes your own decision. Even robots must not be used in decision-making intent.
> 2. *Nihil Nocere* – do not tolerate any harm to users.
> 3. User rights break producer rights.
> 4. Do not build pseudo-autonomous systems that cannot be turned off. Fully autonomous systems should not be allowed.
> 5. The production of self-conscious, autonomously acting robots (if possible) is prohibited (analogous to the chimera ban and human cloning ban in genetic engineering).
> 6. Do not fake a machine as a human subject. A machine must remain machine, imitation and simulation must be always recognizable. It must always be clear to all people involved in human-machine communication that a machine communication partner is a machine.
> 7. If you do not know the question and the purpose of the question, you cannot handle the system response and understand the behaviour of a robot. The context must always be communicated.
> 8. Anyone who invents, who produces, operates or disposes of technology has interests. These interests must be disclosed honestly. [Kornwachs 2021, 42-43]

Taking all the above contributions to heart leads to an unavoidable conclusion: there are no grounds to leave digitalisation out of consideration when emphasising the image of what needs to be upheld as humane. This is all the more valid considering that there is no way to undo digitalisation. Digitalisation is open to, and longs for, humanisation to boost the next step of anthroposociogenesis – a serious attempt of humanising hominisation.

Thus, humanism must not only be specified as convivial and planetary, as eudaimonic and commonist, as done in Chapters 3 and 2, respectively – it can and must also be specified as being digital.

Ultimately, it can be defined:

Techno-social systemism. *Techno-social systemism is that critical techno-social elaboration of* weltanschauung, *that critical conception of the techno-social world and that critical techno-social-scientific way of creating knowledge that devises the criticist Principle of Digital Humanism by applying the criticist Principle of Convivialism to the conditions of the scientific-technological development of digitalisation.*

The **Principle of Digital Humanism** states: there are, regardless of whether or not thematised, associated with the entry of information-technologised social systems into the age of a Great Transformation,

(1) the emergence of digital conscience as information-technologised planetary ethos;

(2) the emergence of digital intelligence as information-technologised planetary agreeability;

(3) the emergence of digital ingeniousness as information-technologised planetary mindfulness;

such that digital conscience, digital intelligence, and digital ingeniousness – thereby specifying convivial transformationism – are techno-social-informational imperatives, instantiated in a hierarchy of information technology levels. At the same time, each is an emergent meta-level on its own, that is, a Third. That Third represents an essential information-technological property of social-informational systems, actualised in values, norms and interests by the interaction networks of the actors as plural Seconds. The participating actors in those Seconds – individual or collective human produsers (producers and users) of information technology – represent single Firsts. Any techno-social-informational system, in the age of a Great Transformation, is exposed to these required specifications of the technologisation of informational self-organisation.

Humanism is instantiated as being digital.

4.5 A Transformation Science for a Shared Humanity

This book and its predecessor outline the key elements of a new paradigm, with its core being the Logic of the Third. This epilogue underlines that this amounts to the nucleus of a transformation science. As a science for, about, and via the third step of anthroposociogenesis, it is a roadmap for guiding, investigating and making ideationally productive the desired and necessary material transformation of social evolution into a future shared humanity.

The Third is everywhere – the meta-level, which can be created by cognisant subjects to understand the metasystemic transitions and suprasystemic transformations that characterise the objective processes and structures in the real world, in the social world, in the social informational world and in the world of social information technology. It is manifest in every realm as an underlying principle that demands attention:

- in emergentism, as the Principle of Unity-through-Diversity;
- in systemism, as the Principle of Self-Organisation;
- in informationism, as the Principle of the Co-Extension of Information and Self-Organisation;
- in sociogenetics, as the Principle of Commonism;
- in noogenetics, as the Principle of Eudaimonism;
- in bifurcationism, as the Principle of Planetarism;
- in transformationism, as the Principle of Convivialism; and
- in techno-social systemism, as the Principle of Digital Humanism.

Each principle fleshes out precedent principles. They yield new qualities that transcend the qualities of the former ones while still being rooted in the former. The structure they build follows the chain of punctuated evolution, unfolding from some agencies as Firsts that interact in some ecology as a Second and can give birth to the structuration of some Third. The agency of this Third can start as another First on a higher level.

Humankind has reached a point of no return. Its actors – individuals or collectives – are potential Firsts that are stuck in a current Second that has yielded a network suffering from morphostasis, stalling the further progress of social morphogenesis[m]. The new formation that is required to overcome the dead-end Second has a structure never seen before – the structure of a social suprasystem that envelops all humanity. This suprasystem is that Third that needs to be visioneered. It needs to be implemented by a different Second, a Second whose actor Firsts conjure that it is apt to bring about the desired Third. These actors can then prove that Third by as-if-actions of different kinds. This, and only this, strategy can set the transformation to the Third in motion. And once the shift to the Third has occurred, the actors become transformed into global netizen Firsts who stabilise the new social entity.

Since this entity is the only alternative that promises the survival and thriving of *Homo sapiens* in an age of self-inflicted existential threats, the transformation science promoted here based on the Logic of the Third is a paradigm shift beyond comparison. It goes far beyond all the wisdom that Copernican revolutions have yet awakened.

Importantly, in line with the deepest knowledge gained so far, any new ideas for saving humanity must by definition be an extension, a concretisation and a consolidation of the framework presented here. This calls for all further investigations into the Great Transformation to make methodological use of the tenets outlined here.

[m] As Margaret Archer terms the forming of societies to new societal formations.

Bibliography

Acton, J.M. (2021). The U.S. exit from the Anti-Ballistic Missile Treaty has fueled a new arms race, *Carnegie Endowment for International Peace*, https://carnegieendowment.org/2021/12/13/u.s.-exit-from-anti-ballistic-missile-treaty-has-fueled-new-arms-race-pub-85977 (retrieved 31.01.2022).

Adloff, F., Costa, S. (2020). Konvivialismus 2.0: Ein Nachwort, (ed.) Die konvivialistische Internationale, *Das zweite konvivialistische Manifest, Für eine post-neoliberale Welt* (transcript, Bielefeld), pp. 119-140.

Alexander, J.C. (1995). *Fin de Siecle social theory – relativism, reduction, and the problem of reason* (Verso, London).

Anders. G. (2016). On promethean shame, ed. Müller, C.J., *Prometheanism – technology, digital culture and human obsolescence* (Rowman and Littlefield, London), pp. 29-95.

Archer, M.S. (1995). *Realist social theory – the morphogenetic approach* (Cambridge University Press, Cambridge).

Archer, M.S. (2003). *Structure, agency and the internal conversation* (Cambridge University Press, Cambridge).

Archer, M.S. (2007). *Making our way through the world – human reflexivity and social mobility* (Cambridge University Press, Cambridge).

Archer, M.S. (2010). *Conversations about reflexivity* (Routledge, London).

Archer, M.S. (2012). *The reflexive imperative in late modernity* (Cambridge University Press, Cambridge).

Archer, M.S. (2015). *Morphogenesis and human flourishing* (Springer, Cham).

Artigiani, R. (1991). Social evolution: a nonequilibrium systems model, ed. Laszlo, E., *The new evolutionary paradigm* (Gordon and Breach, New York), pp. 93-131.

Ashby, W.R. (1956). *An Introduction to Cybernetics* (Wiley, New York).

Atomic Heritage Foundation (2017). The Human Computers of Los Alamos, https://www.atomicheritage.org/history/human-computers-los-alamos (retrieved 31.01.2022).

Bammé, A. (2011). *Homo occidentalis: von der Anschauung zur Bemächtigung der Welt* (Velbrück, Weilerswist-Metternich)

Barad, K. (2007). *Meeting the universe halfway: quantum physics and the entanglement of matter and meaning* (Duke University Press, Durham).

Barad, K. (2012). *Agentieller Realismus: über die Bedeutung materiell-diskursiver Praktiken* (Suhrkamp, Berlin).

Barthlott, W., Linsenmair, K.E., Porembski, S. (eds.) (2009). *Biodiversity: structure and function, vol. ii* (EOLSS, Oxford).

Baudrillard, J. (1995). *Simulacra and simulation* (The University of Michigan Press, Ann Arbor).

Becker, J. (2002). *Information und Gesellschaft* (Springer, Wien).

Becker, J., Beham, M. (2006). *Operation Balkan – Werbung für Krieg und Tod* (Nomos, Baden-Baden).

Beniger, J.R. (1986). *The control revolution* (Harvard University Press, Cambridge).

Berger, J. (2021). Wenn westlicher Qualitätsjournalismus, Propaganda und Infokrieg Hand in Hand gehen, *Nachdenkseiten*, https://www.nachdenkseiten.de/?p=70340 (retrieved 31.01.2022).

Bertalanffy, L.v. (1928). *Nikolaus von Kues* (Georg Müller, München).

Bertalanffy, L.v. (1932). *Theoretische Biologie*. Band 1 (Gebrüder Borntraeger, Berlin).

Bertalanffy, L.v. (1950). An outline of General System Theory, *British Journal for the Philosophy of Science*, vol.1, no. 2, pp. 134-165.

Bertalanffy, L.v. (1953). Philosophy of science in scientific education, *The Scientific Monthly*, 77, pp. 233-239.

Bishop, S., Helbing, D. (2012). FutureICT Flagship Proposal, Summary version of the full 620 page final proposal (FutureICT Consortium), http://www.hofkirchner.uti.at/wp-content/uploads/2013/01/SUMMARY20FuturICT20Proposal20820Nov.pdf (retrieved 31.01.2022).

Bloch, E. (1985). *Das Prinzip Hoffnung*, 3 vols. (Suhrkamp, Frankfurt am Main).

Bloch, E. (1986). *The Principle of Hope* (MIT Press, Cambridge, Massachusetts).

Bornemann, E. (1975). *Das Patriarchat* (S. Fischer, Frankfurt am Main).

Borsche, T. (2017). Begriffe – die Urformen menschlicher Artefakte, ed. Franz, J.H., Berr, K., *Welt der Artefakte* (Frank & Timme, Berlin), pp. 29-42.

Bowles, S., Gintis, H. (2011). *A cooperative species. Human reciprocity and its evolution* (Princeton University Press, Princeton).

Bradley, G. (2006). *Social and community informatics. Humans on the net* (Routledge, London).

Braga, A., Logan, R.K. (2017). The emperor of strong AI has no clothes: limits to artificial intelligence, *Information*, 8, 4, https://www.mdpi.com/2078-2489/8/4/156 (retrieved 31.01.2022).

Braga, A., Logan, R.K. (2021). The singularity hoax: why computers will never be more intelligent than humans, eds. Hofkirchner, W., Kreowski, H.-J., *Transhumanism: the proper guide to a posthuman condition or a dangerous idea?* (Springer, Cham), pp. 133-140.

Brand, U., Wissen, M. (2021). *The imperial mode of living – Everyday life and the ecological crisis of capitalism* (Verso, London and New York).

Brenner, J.E. (2008). *Logic in reality* (Springer, Dordrecht).

Brenner, J.E., Igamberdiev, A.U. (2021). *Philosophy in reality: a new book of changes* (Springer, Cham),

Briley, R. (2006). "Woody Sez": Woody Guthrie, the People's Daily World, and indigenous radicalism, *California History*, 84 (1), pp. 30-43.

Bruns, A. (2006). Towards produsage: Futures for user-led content production, eds. Sudweeks, F., Hrachovec, H., Ess, C., *Proceedings cultural attitudes towards communication and technology 2006* (Tartu), pp. 275-284.

Buber, M. (1923). *Ich und du* (Insel-Verlag, Leipzig).

Buckley, W.F. (1967). *Sociology and modern systems theory* (Prentice Hall, Englewood Cliffs).

Bunge, M. (2003). *Emergence and convergence* (University of Toronto Press, Toronto).

Bunge, M. (2012). *Evaluating philosophies* (Springer, Dordrecht).

Burgin, M., Hofkirchner, W. (2017). Introduction: Omnipresence of information as the incentive for transdisciplinarity, eds. Burgin, M., Hofkirchner, W., *Information Studies and the Quest for Transdisciplinarity – Unity through Diversity* (World Scientific, Singapore), pp. 1-7.

Buro, A., Singe, M. (2009). Expansion und Eskalation: 60 Jahre NATO, *Blätter*, 4, pp.53-63.

Bulletin of the Atomic Scientists, https://thebulletin.org/doomsday-clock/ (retrieved 31.01.2022).

Bühl, A. (1997). *Die virtuelle Gesellschaft* (Westdeutscher Verlag, Wiesbaden).

Capurro, R. (2012). Toward a comparative theory of agents, *AI & Society*, 27, 4, pp. 479-488.

Carson, R. (1962). *Silent spring* (Houghton Mifflin, Boston).

Castells, M. (2010a). *The information age: economy, society, and culture, 2: The Power of Identity* (Blackwell, Malden, MA).

Castells, M. (2010b). *The information age: economy, society, and culture, 3: End of Millennium* (Blackwell, Malden, MA).

Castells, M. (2011). *The information age: economy, society, and culture, 1: The rise of the network society* (Blackwell, Malden, MA).

Castells, M. (2013). *Communication power* (Oxford University Press, New York).

Chomsky, N. {2021). *Rebellion oder Untergang! Ein Aufsruf zu globalem Ungehorsam zur Rettung unserer Zvilisation* (Westend, Frankfurt am Main).

Coenen, C., Hofkirchner, W., Díaz Nafría, J.M. (2012a). New ICTs and social media in political protest and social change, *IRIE*, 18, pp. 2-8.

Coenen, C., Hofkirchner, W., Díaz Nafría, J.M. (eds.) (2012b). New ICTs and social media: revolution, counter-revolution and social change, *IRIE*, 18.

Collier, J., Stingl, M. (2020). Evolutionary moral realism (Routledge, Abingdon and New York).

Convivialist International (2020). The Second Convivialist Manifesto – Towards a Post-Neoliberal World. *Civic Sociology*, https://online.ucpress.edu/cs/article/1/1/12721/112920/THE-SECOND-CONVIVIALIST-MANIFESTO-Towards-a-Post (retrieved 31.01.2022).

Convivialist Manifesto (2014). *A declaration of interdependence* (Global Dialogues, Duisburg), doi: 10.14282/2198-0411-GD-3.

Corning, P. (1983). *The Synergism Hypothesis – a theory of progressive evolution* (McGraw-Hill, New York).

Corning, P. (2003). *Nature's magic – synergy in evolution and the fate of humankind* (Cambridge University Press, Cambridge).

Costanza, R. (2003). A vision of the future science: reintegrating the study of human and the rest of nature, *Futures*, 35, pp. 651-671.

Crutzen, P.J., Stoermer, E.F. (2000). The "Anthropocene", *IGBP Global Change Newsletter*, 41, May 2000, pp. 17-18.

Cumings, B. (2004). Why did Truman really fire MacArthur? … The obscure history of nuclear weapons and the Korean War provides the answer, *Le Monde Diplomatique*, December 2004, hnn.us/articles/9245.html.

Curtis, N. (2013). *Idiotism: capitalism and the privatisation of life* (PlutoPress, London).

Davidson, M. (1983). *Uncommon sense. The life and thought of Ludwig von Bertalanffy, the father of general system theory* (Tarcher, Los Angeles).

Davidson, M. (2005), ed. Hofkirchner, W., *Querdenken! Leben und Werk Ludwig von Bertalanffys* (Peter Lang, Frankfurt am Main).

Debord, G. (1967). *La société du spectacle* (Buchet, Chastel, Paris).

Deutsches Referenzzentrum für Ethik in den Biowissenschaften. *Anthroporelational.* Available online: https://www.drze.de/im-blickpunkt/biodiversitaet/module/anthroporelational (retrieved 25 August 2021).

de Kerckhove, D. (1998). *Connected intelligence* (KoganPage, London).

Dietschy, B., Zeilinger, D., Zimmermann, R.E. (2012). *Bloch-Wörterbuch* (De Gruyter, Berlin).

Djelic, M., Quack, S. (2010). Transnational communities and governance, eds. Djelic, M., Quack, S., *Transnational Communities* (Cambridge University Press, Cambridge) pp. 3-36.

Dodig-Crnkovic, G., Hofkirchner, W. (2011). Floridi's "Open problems in Philosophy of Information", ten years later, *Information*, 2, pp. 327-359.

Donati, P. (2011). *Relational Sociology – a new paradigm for the social sciences* (Routledge, London).

Donati, P. (2014). Relational goods and their subjects: the ferment of a new civil society and civil democracy, *Recerca*, 14, pp. 19-46.

Donati, P., Archer, M.S. (2015). *The relational subject* (Cambridge University Press, Cambridge).

Eberl, J.-M., Lebernegg, N.S. (2021). Corona-Demonstratnt*innen: Rechts, wissenschaftsfeindlich und esoterisch, https://viecer.univie.ac.at/en/projects-andcooperations/austrian-corona-panel-project/corona-blog/corona-blog-beitraege/blog138/ (retrieved 31.01.2022).

Eberl, J.-M., Partheymüller, J., Paul, K.T. (2021). Impfbereitschaft, impfpolitische Maßnahmen und Kinderimpfung, https://viecer.univie.ac.at/corona-blog/corona-blog-beitraege/blog129/ (retrieved 31.01.2022).

Eisler, R. (1987). *The Chalice and the Blade: Our History, Our Future* (Harper and Row, San Francisco).

Eller, C. (2000). *The Myth of Matriarchal Prehistory* (Beacon Press, Boston).

European Commission, Directorate-General for Employment, Industrial Relations and Social Affairs (ed.) (1997). *Building the European information society for us all, Final Policy Report of the high-level expert group* (Office for Official Publications of the European Communities, Luxembourg).

Fischer-Kowalski, M., Haberl, H. (1993). *Metabolism and colonisation: modes of production and the physical exchange between societies and nature* (IFF Soziale Ökologie, Wien).

Fleissner, P., Hofkirchner, W. (1998). The making of the information society: driving forces, 'Leitbilder' and the imperative for survival, *BioSystems*, 46, pp.201-207.

Floridi, L. (2007). A look into the future impact of ICTs on our lives, *The Information Society*, 23, 1, pp. 59-64.

Floridi, L. (2010). *Information. A very short introduction* (Oxford University Press, Oxford).

Floridi, L. (2013). *The ethics of information* (Oxford University Press, Oxford).

Floridi, L., Cowls, J., Beltrametti, M., Chatila, R., Chazerand, P., Dignum, V., Luetge, C., Madelin, R., Pagallo, U., Rossi, F., Schafer, B., Valcke, P., Vayena, E. (2018). AI4People – An ethical framework for a Good AI Society: opportunities, risks, principles, and recommendations, *Minds and Machines*, 28, pp. 689-707.

Franck, G. (2010). *Ökonomie der Aufmerksamkeit: ein Entwurf* (Hanser, München, Wien).

Franta, B. (2018). On its 100th birthday in 1959, Edward Teller warned the oil industry about global warming, *The Guardian*, 1 Jan 2018, https://www.theguardian.com/environment/climate-consensus-97-per-cent/2018/jan/01/on-its-hundredth-birthday-in-1959-edward-teller-warned-the-oil-industry-about-global-warming (retrieved 03.01.2022).

Fuchs, C., Hofkirchner, W. (2009). Autopoiesis and critical social systems theory, eds. Magalhães, R., Sanchez, R., *Autopoiesis in organization theory and practice* (Emerald, Bingley), pp. 11-129.

Fuchs-Kittowski, K. (1997). Information – neither matter nor mind, on the essence and on the evolutionary stage concept of information, *World Futures*, vol. 49/50, pp.551-570.

Gartner Research (2015). Digital Business: Digital Humanism Makes People Better, Not Technology Better, 22 April 2015, https://www.gartner.com/en/documents/3035017/digital-humanism-makes-people-better-not-technology-better (retrieved 24.01.2022).

Giddens, A. (1984). *The constitution of society* (Polity Press, Cambridge).

Gintis, H., Helbing, D. (2015). Homo socialis: an analytical core for sociological theory, *Review of Behavioral Economics*, 2, pp. 1-59.

Gonnermann, B., Mechtersheimer, A. (eds.) (1990). *Verwundbarer Frieden, Zwang zu gemeinsamer Sicherheit für die Industriegesellschaften Europas* (Brandenburgisches Verlagshaus, Berlin).

Graeber, D., Wengrow, D. (2021). *The dawn of everything: a new history of humanity* (Allen Lane, Penguin, London, New York).

Grimalda, G. (2016). The possibilities of global we-identities, eds. Messner, D., Weinlich, S., *Global Cooperation and the Human Factor in International Relations* (Routledge, London). pp. 201-224.

Gould, S.J. (2002). *The structure of evolutionary theory* (Belknap, Cambridge).

Grossegger, B. (2021). "Viele sorgen sich um ihre Zukunftschancen", https://www.migros.ch/de/Magazin/2021/03/beate-grossegger-jugendforscherin. html (retrieved 31.01.2022).

Haftor, D.M., Mirijamdotter, A. (eds.) (2011). *Information and communication technologies, society and human beings: Theory and framework* (Information Science Reference, Hershey, Pennsylvania).

Hancock, P.A. (2009). *Mind, machine and morality: toward a philosophy of human-technology symbiosis* (Ashgate, Farnham).

Harari, Y.N. (2016). *Homo deus – a brief history of tomorrow* (Harvill Secker, London).

Haug, W.F. (1989). *Gorbatschow, Versuch über den Zusammenhang seiner Gedanken* (Argument-Verlag, Hamburg).

Haug, W.F. (2004). *Historisch.Kritisches Wörterbuch des Marxismus*, vol. 6/II (Argument, Berlin).

Haefner, K. (1992). Information Processing at the Sociotechnical Level, ed. Haefner, K., *Evolution of Information Processing Systems* (Springer, Berlin etc.), pp. 307-319.

Heinz, J., Ogris, G. (2021). *Freiheitsindex Österreich 2020 Inkl. Follow-Up 2021* (SORA, Wien). https://www.sora.at/fileadmin/downloads/projekte/2021_Bericht_Freiheitsindex_Oesterreich.pdf (retrieved 31.01.2022).

Heinzlmaier, B. (2012). Keine Mission, keine Vision, keine Revolution? Die postmoderne Jugend zwischen Pragmatismus und Idealismus, http://jugendkultur.at/wp-content/uploads/keine_mission_heinzlmaier_2012.pdf (retrieved 31.01.2022).

Heinzlmaier, B. (2021). *Generation Corona: Über das Erwachsenwerden in einer gespaltenen Gesellschaft* (Hirnkost, Berlin).

Heylighen, F. (2015). Return to Eden? Promises and perils on the road to a global superintelligence, eds. Goertzel, B., Goertzel, T., *The end of the beginning: life, society and economy on the brink of the singularity*, pp. 243-306 (Humanity+ press, Los Angeles).

Heylighen, F. (2016). A brain in a vat cannot break out: why the singularity must be extended, embedded and embodied, ed. Awret, U., *The Singularity: could artificial intelligence really out-think us (and would we want it to)?*, vol. 19, pp. 126-142 (Andrews UK Limited, Luton).

Hilty, L.M., Aebischer, B. (eds.) (2015). *ICT innovatons for sustainability*, Advances in Intelligent Systems and Computing 310 (Springer, Switzerland).

Hofkirchner, W. (1996). Computer Integrated Warfare: Die Flexibilisierung der Destruktion. Fleissner, P., Hofkirchner, W., Müller, H., Pohl, M., Stary. C., *Der Mensch lebt nicht vom Bit allein* (Peter Lang, Frankfurt), pp. 205-239.

Hofkirchner, W. (1998). Emergence and the logic of explanation: An argument for the unity of science, *Acta Polytech Scand Math Comput Manag Eng Ser*, 91, pp. 23-30.

Hofkirchner, W. (2003). Homo creator in einem schöpferischen Universum, eds. Maurer, M., Höll, O., *Natura als Politikum* (RLI-Verlag, Wien), pp. 371-392.

Hofkirchner, W. (2006). Towards a post-Luhmannian social systems view, *International Sociological Association: Proceedings of the XVI World Congress of Sociology: The Quality of Social Existence in a Globalizing World, Durban* (CD-ROM), pp. 1-20, http://www.hofkirchner.uti.at/icts-wh-profile/pdf1453.pdf (retrieved 31.01.2022).

Hofkirchner, W. (2007). A critical social systems view of the internet, *Philosophy of the Social Sciences*, 37, 4, pp. 471-500.

Hofkirchner, W. (2010a). A taxonomy of theories about ICTs and Society, *triple-c*, 8, 2, pp. 171-176.

Hofkirchner, W. (2010b). How to design the Infosphere: the Fourth Revolution, the management of the life cycle of information, and information ethics as a macroethics, *Know Techn Pol*, 23, pp. 177-192.

Hofkirchner, W. (2011a). Information and communication technologies for the Good Society, Haftor, D.M., Mirijamdotter, A. (eds.), *Information and communication technologies, society and human beings: Theory and framework* (Information Science Reference, Hershey, Pennsylvania), pp. 434-443.

Hofkirchner, W. (2011b). Does computing embrace self-organisation? eds. Dodig-Crnkovic, G., Burgin, M., *Information and computation* (World Scientific, Singapore), pp. 185-202.

Hofkirchner, W. (2012). Sustainability and self-organisation: sustainability in the perspective of complexity and systems science and ethical considerations, eds. Nishigaki, T., Takenouchi, T., *Information Ethics: The Future of the Humanities* (V2 Solution Publisher, Nagoya City).

Hofkirchner, W. (2013a). *Emergent Information. A Unified Theory of Information Framework* (World Scientific, Singapore).

Hofkirchner, W. (2013b). Self-organisation as the mechanism of development and evolution of social systems, ed. Archer, M.S., *Social morphogenesis* (Springer, Cham), pp. 125-143.

Hofkirchner, W. (2014a). Potentials and risks for creating a global sustainable information society, ed. Fuchs, C. and Sandoval, M., *Critique, social media and the information society* (Routledge, London), pp. 66-75.

Hofkirchner, W. (2014b). The commons from a critical social systems perspective, *Recerca*, 14, pp. 73-91.

Hofkirchner, W. (2014c). On the validity of describing 'morphogenic society' as a system and justifiability of thinking about it as a social formation, ed. Archer, M.S., *Late modernity: Trajectories towards morphogenic society* (Springer, Cham), pp.119-141.

Hofkirchner, W. (2014d). Idiotism and the Logic of the Third, ed. Lakitsch, M., *Political power reconsidered* (Lit Verlag, Wien), pp. 55-75.

Hofkirchner, W. (2015). "Mechanisms" at work in Information Society, ed. Archer, M.S., *General mechanisms transforming the social order* (Springer, Cham), pp. 95-112.

Hofkirchner, W. (2016). Ethics from systems: origin, development and current state of normativity, ed. Archer, M.S., *Morphogenesis and the crisis of normativity* (Springer, Cham), pp. 279-295.

Hofkirchner, W. (2017a). Information for a Global Sustainable Information Society, ed. Hofkirchner, W. and Burgin, M., *The future information society, Social and technological problems* (World Scientific, Singapore). pp. 11-33.

Hofkirchner, W. (2017b). Creating common good: The Global Sustainable Information Society as the good society, ed. Archer, M.S., *Morphogenesis and human flourishing* (Springer, Cham), pp. 277-296.

Hofkirchner, W. (2017c). Transdisciplinarity needs systemism, *Systems*, 5, 15, 1-15,11 (2017a), https://doi.org/10.3390/systems5010015 (retrieved 31.01.2022).

Hofkirchner, W. (2019). Social relations: Building on Ludwig von Bertalanffy, *Syst Res Behav Sci.*, vol. 36, May/June 2019, pp. 263-273, https://onlinelibrary.wiley.com/doi/10.1002/sres.2594 (retrieved 31.01.2022).

Hofkirchner, W. (2020a). Intelligence, Artificial Intelligence and Wisdom in the Global Sustainable Information Society, *proceedings*, 2020, 47, 39, pp. 1-4, doi:10.3390/proceedings2020047039.

Hofkirchner, W. (2020b). Taking the perspective of the Third. A contribution to the origins of systems thinking, *proceedings*, 2020, 47, 8, pp. 1-4, doi:10.3390/proceedings2020047008.

Hofkirchner, W. (2020c). Blurring of the human and the artificial: a conceptual clarification, proceedings, 2020, 47, 7, pp. 1-3, doi:10.3390/proceedings2020047007.

Hofkirchner, W. (2020d). A paradigm shift for the Great Bifurcation, *Biosystems*, 197, pp. 1-7, https://doi.org/10.1016/j.biosystems.2020.104193 (retrieved 31.01.2022).

Hofkirchner, W. (2021a). Promethean shame revisited: a praxio-onto-epistemological analysis of cyber futures, eds. Hofkirchner, W., Kreowski, H.-J., *Transhumanism: the proper guide to a posthuman condition or a dangerous idea?* (Springer, Cham), pp. 185-206.

Hofkirchner, W. (2021b). Digital humanism: epistemological, ontological and praxiological foundations, ed. Verdegem, P., *AI for everyone?, Critical perspectives* (University of Westminster Press, London), pp. 33-47.

Hofkirchner, W. (2021c). Zur Konzeptualisierung von Wahrheit aus der Sicht der Praxio-Onto-Epistemologie II, *Signifikant, Jahrbuch für Strukturwandel und Diskurs*, 3, 2020, pp. 167-179.

Hofkirchner, W. (2022). Artificial Intelligence: "Machines of loving grace" or "Tools for conviviality"?, *Signifikant, Jahrbuch für Strukturwandel und Diskurs*, 4, in print.

Hofkirchner, W., Díaz-Nafría, J.M., Crowley, P., Graf, W., Kramer, G., Kreowski, H.-J., Wintersteiner, W. (2019). ICTs connecting Global Citizens, Global Dialogue and Global Governance – a call for needful designs, *CCIS*, vol. 1051, pp. 453-468.

Hofkirchner, W., Kreowski, H.-J. (2021). *Transhumanism: the proper guide to a posthuman condition or a dangerous idea?* (Springer, Cham).

Hofkirchner, W., Kreowski, H.-J. (2022). Digital Humanism: how to shape digitalisation in the age of global challenges?, *proceedings*, 2022, 68.

Hofkirchner, W., Fuchs, C., Raffl, C., Schafranek, M., Sandoval, M., Bichler, R. (2007). ICTs and Society: The Salzburg Approach. Towards a theory for, about, and by means of the information society, Research Paper No. 3 (ICT&S Center, Salzburg), http://www.hofkirchner.uti.at/icts-wh-profile/pdf1490.pdf {retrieved 08.07.2020).

Holling, C.S. (1973). Resilience and stability of ecological systems, *Annual Review of Ecology and Systematics*, 4, pp.1-23.

Holling, E., Kempin, P. (1989). *Identität, Geist und Maschine* (Rowohlt, Reinbek bei Hamburg).

Holzkamp, K. (1983). *Grundlegung der Psychologie* (Campus, Frankfurt am Main).

Horkheimer, M., Adorno, T.W. (2016). *Dialektik der Aufklärung* (Fischer, Frankfurt/M.)

Hörz, H.E., Hörz, H. (2013). *Ist Egoismus unmoralisch? Grundzüge einer neomodernen Ethik* (Trafo, Berlin).

Illich, I. (1973). *Tools for conviviality* (Marion Boyars, London).

Illich, I. (1975). *Selbstbegrenzung, Eine politische Kritik der Technik* (Rowohlt, Reinbek bei Hamburg)

Jantsch, E. (1987). Erkenntnistheoretische Aspekte der Selbstorganisation natürlicher Systeme, ed. Schmidt, S. J., *Der Diskurs des Radikalen Konstruktivismus* (Suhrkamp, Frankfurt am Main,) pp. 159-191.

Jonas, H. (1979). *Das Prinzip Verantwortung: Versuch einer Ethik für die technologische Zivilisation* (Insel-Verlag, Frankfurt am Main).

Jonas, H. (1984). *The imperative of responsibility: in search of an ethics of the technological age* (University of Chicago: Chicago).

jugendkultur (n.a.). Buch: Generation Corona, https://jugendkultur.at/buch-generation-corona/#more-14988 (retrieved 31.01.2022).

Jungen, O. (2017). Silicon Valley Simuliert nur, *Frankfurter Allgemeine*, 28 January 2017. https://www.faz.net/aktuell/feuilleton/debatten/oswald-wiener-und-die-kuenstliche-intelligenz-14770061.html (retrieved 24.01.2020).

Kaldor, M. (1982). *The baroque arsenal* (Hill and Wang, New York).

Kim, J., Oki, T. (2011). Visioneering: an essential framework in sustainability science, *Sustain Sci*, 6, pp. 247-251.

Klaus, G., Buhr, M. (1974). *Philosophisches Wörterbuch* (VEB Bibliographisches Institut, Leipzig).

Knies, G., Gonnermann, B., Schmidt-Eenboom (eds.) (1990). *Betriebsbedingung Frieden, Herausforderungen der Hochtechnologie-Zivilisation für eine nachmilitärische Ära* (Brandenburgisches Verlagshaus, Berlin).

Kornwachs, K. (2021). Transhumanism as a derailed anthropology, ed. Hofkirchner, W., Kreowski, H.-J., *Transhumanism: the proper guide to a posthuman condition or a dangerous idea?* (Springer, Cham), pp. 21-47.

Krämer, S. (1988). *Symbolische Maschinen* (Wissenschaftliche Buchgesellschaft, Darmstadt).

Kreisberg, J.C. (1995). An obscure Jesuit priest, Pierre Teilhard de Chardin, set down the philosophical framework for planetary, net-based consciousness 50 years ago, *WIRED*, 3, 6, pp. 108-113.

Krüger, P. (1981). *Wladimir Iwanowitsch Wernadskij* (Teubner. Leipzig).

Krüger, U. (2019). *Meinungsmacht: Der Einfluss von Eliten auf Leitmedien – eine kritische Netzwerkanalyse* (Herbert von Halem, Köln).

Kuhn, T.S. (1962). *The structure of scientific revolutions* (University of Chicago Press, Chicago).

Küng, H. (1997). *Weltethos für Weltpolitik und Weltwirtschaft* (Piper, München).

LANL (2017). How did the Manhattan Project impact computing?, https://www.lanl.gov/museum/discover/_docs/manhattan-project-computing.pdf (retrieved 31.01.2022).

Latour, B. (2006). *Reassembling the social: an introduction to Actor-Network-Theory* (Oxford University Press, Oxford).

Laszlo, E. (1986). *La grande bifurcation* (Tacor International, Versailles).

Laszlo, E. (1990). *La gran bifurcación* (Gedisa Editorial, Barcelona).

Laszlo, E. (2001). *Macroshift – Navigating the transformation to a sustainable world* (Berrett-Koehler, San Francisco).

Laszlo, E. (2010). *The chaos point: the world at the crossroads* (Piatkus, London).

Latour, B. (2005). *Reassembling the social. An introduction into Actor-Network-Theory* (Oxford University Press, Oxford).

Lawson, T. (2013). Emergence and morphogenesis – causal reduction and downward causation? ed. Archer, M.S., *Social morphogenesis* (Springer, Cham), pp. 61-84.

Lawson, T. (2017). Eudaimonic bubbles, social change and the NHS, ed. Archer, M.S., *Morphogenesis and human flourishing* (Springer, Cham), pp.239-260.

Levy, P. (1997). *Collective Intelligence* (Plenium Trade, New York etc.).

Lévy, P. (2001). *Cyberculture* (University of Minnesota Press, Minneapolis etc.).

Lieber, K.A., Press, D.G. (2006). The rise of U.S. nuclear primacy, *Foreign Affairs*, 2, pp. 42-54.

Linebaugh, P. (2014). *Stop, thief! The commons, enclosure, and resistance* (PM Press, Oakland).

Logan, Robert K. (2007). *The extended mind: the emergence of language, the human mind and culture* (University of Toronto Press, Toronto).

Lovelock, J. (1979). *Gaia: a new look at life on earth* (Oxford University Press, Oxford).

Lovelock, J. (2019). *Novacene: the coming age of hyperintelligence* (Allen Lane, London).

Luhmann, N. (2001). *Die Gesellschaft der Gesellschaft*, 2 vols. (Suhrkamp, Frankfurt am Main).

Lynas, M., Houlton, B.Z., Perry, S. (2021). Greater than 99% consensus on human caused climate change in the peer-reviewed scientific literature, Environmental Research Letters, 16, 11, https://iopscience.iop.org/article/10.1088/1748-9326/ac2966 (retrieved 31.01.2022).

Madrigal, A.C. (2011). Weekend Poem: All Watched Over by Machines of Loving Grace, *The Atlantic*, https://www.theatlantic.com/technology/archive/2011/09/weekend-poem-all-watched-over-by-machines-of-loving-grace/245251/ (retrieved 11.08.2020).

Marotzke, J. (2022). Beim CO2-Budget muss die Buchhaltung stimmen, *Der Standard*, 5/6 January, p. 16.

Marx, K. (1843). A Contribution to the Critique of Hegel's Philosophy of Right, Introduction, http://www.marxists.org/archive/marx/works/1843/critique-hpr/intro.htm (retrieved 31.01.2022).

Marx, K. (1852). The Eighteenth Brumaire of Louis Bonaparte. http://www.marxists.org/archive/marx/works/1852/18th-brumaire/ch01.htm (retrieved 31.01.2022).

Maser, C. (1999). *Vision and leadership in sustainable development* (CRC, West Palm Beach).

Matlock, J.F. (2021). Ukraine: tragedy of a nation divided, *KRASNO Weekly Spotlight*, 2

Maturana, H.R., Varela, F. (1980). *Autopoiesis and cognition* (Reidel, Dordrecht).

Mead, G.H. (1934). *Mind, self, and society* (University of Chicago Press, Chicago).

Mearsheimer, J.J. (2014). Why the Ukraine crisis is the West's Fault. The liberal delusions that provoked Putin, *Foreign Affairs*, 93, 5, pp. 77-89.

Menasse, R. (2012) *Der Europäische Landbote* (Zsolnay, Wien).

Merton, R.K. (1973). *The sociology of science. Theoretical and empirical investigations* (University of Chicago Press, Chicago).

Messner, D., Weinlich, S. (2016). The evolution of human cooperation, eds. Messner, D., Weinlich, S., *Global Cooperation and the Human Factor in International Relations* (Routledge, London) pp. 3-46.

Mitchell, M. (2019). *Artificial Intelligence: a guide for thinking humans* (Farrar, Straus and Giroux, New York).

Morin, E. (1979). *Le paradigm perdu* (Seuil, Paris).

Morin, E. (1992). *The nature of nature* (Lang, New York).

Morin, E. (1999). *Homeland Earth. A manifesto for the new millennium* (Hampton Press, Creskill).

Morin, E. (2012). *Der Weg, Für die Zukunft der Menschheit* (Krämer, Hamburg).

Morin, E. (2021). Abenteuer Mensch, *Freitag*. 2021, 28. Available online: https://www.freitag.de/autoren/the-guardian/abenteuer-mensch (retrieved 25.08.2021).

Mouffe, C. (2013). *Agonistics. Thinking the world politically* (Verso, London).

Mouzelis, N. (1995). *Sociological theory: what went wrong?* (Routledge, London).

Munroe, M. (2003). *The principles and power of vision: keys to achieving personal and corporate destiny* (Whitaker House, New Kensington).

Müller, A. (2019). *Glaube wenig – Hinterfrage alles – Denke selbst. Wie man Manipulationen durchschaut* (Westend, Frankfurt am Main).

Negroponte, N. (1995). *Being digital* (Knopf, New York).

Neumann, I. (2016) Diplomatic cooperation, eds. Messner, D., Weinlich, S., *Global Cooperation and the Human Factor in International Relations* (Routledge, London), pp. 225-245.

Nida-Rümelin, J, Weidenfeld, N. (2018). *Digitaler Humanismus. Eine Ethik für das Zeitalter der Künstlichen Intelligenz* (Piper, München).

Niedenzu, H.-J. (2012). *Soziogenese der Normativität* (Velbrück, Weilerswist).

Nora, S., Minc, A. (1978). *L'informatisation de la Société: rapport à M. le Président de la République* (La Documentation française, Paris).

Nowak, M., Highfield, R. (2011). *Super co-operators. Evolution, Altruism and human behaviour or why we need each other to succeed* (Canongate, Edinburgh).

Nowotny, H. (2021). *In AI we trust: power, illusion and control of predictive algorithms* (polity, Cambridge).

Obrecht, W. (2005). Der emergentistische Systemismus Mario Bunges und das Systemtheoretische Paradigma der Sozialarbeitswissenschaft und der Sozialen Arbeit (SPSA), http://www.sozialarbeit.ch/kurzinterviews/werner_obrecht.htm (retrieved 07.07.2020).

OECD (2010). *ICTs for Development: improving policy coherence* (OECD iLibrary).

Oeser, E. (1992). Mega-Evolution of Information Processing Systems, ed. Haefner, K., *Evolution of Information Processing Systems* (Springer, Berlin etc.), pp. 103-111.

Ord, T. (2021). *The precipice: existential risk and the future of humanity* (Bloomsbury, London).

OXFAM International (2021). *The inequality virus*, report, https://oxfamilibrary.openrepository.com/bitstream/handle/10546/621149/bp-the-inequality-virus-summ-250121-en.pdf (retrieved 31.01.2022).

Paech, N. (2014). Wem gehört die Krim? Die Krimkrise und das Völkerrecht, ed. Strutyinksi, P., *Ein Spiel mit dem Feuer, die Ukraine, Russland und der Westen* (Papyrossa, Köln), pp. 54-64.

Paech, N. (2019). "Not one inch eastwards", Streit um NATO-Osterweiterung, *Z*, December, pp. 161-165.

Pagel, M. (2012). *Wired for culture. Origins of the human social mind* (Norton, New York).

Peirce, C.S. (2000). *Semiotische Schriften*, 3 vols. (Suhrkamp, Frankfurt).

Peschl, M.F., Fundneider, T. (2012). Spaces enabling game-changing and sustaining innovations, *OTSC* 9 (1), pp. 41-61.

Piketty, T. (2014). *Capital in the twenty-first century* (Belknap Press of Harvard University Press, Cambridge, Massachusetts).

Porpora, D.V. (2001). *Landscapes of the soul: the loss of moral meaning in american life* (Oxford University Press, New York).

Porpora, D.V., Shumar, W. (2010). Self talk and self reflection, ed. Archer, M.S., *Conversations about reflexivity* (Routledge, London), pp. 206-220.

Pörksen, B. (2018). *Die große Gereiztheit: Wege aus der kollektiven Erregung* (Hanser, München).

Precht, R.D. (2021). *Von der Pflicht* (Goldmann, München).

Reckwitz, A. (1997). *Struktur – Zur sozialwissenschaftlichen Analyse von Regeln und Regelmäßigkeiten* (Westdeutscher Verlag, Opladen).

Reuters Institute for the Study of Journalism (2017). Digital News Report, http://www.digitalnewsreport.org (retrieved 31.01.2022).

Reuters Institute for the Study of Journalism (2019). Digital News Report, http://www.digitalnewsreport.org (retrieved 31.01.2022).

Rheingold, H. (1993). *The virtual community: homesteading on the electronic frontier* (Addison-Wesley, Reading, MA).

Rheingold, H. (2000). *Tools for thought: the history and future of mind-expanding technology* (MIT Press, Cambridge).

Rheingold, H. (2002). *Smart mobs: the next social revolution* (Perseus, Cambridge, MA).

Rheingold, H. (2005). *Technologies of cooperation* (Institute for the Future, Palo Alto), https://www.rheingold.com/cooperation/Technology_of_cooperation.pdf (retrieved 31.01.2022).

Ritz, H. (2008). Die Welt als Schachbrett. Der neue Kalte Krieg des Obama-Beraters Zbigniew Brzezinski, *Blätter*, 7, pp. 53-69.

Robertson, R. (1992). *Globalization* (Sage, London)

Ropohl, G. (2012). *Allgemeine Systemtheorie, Einführung in transdisziplinäres Denken* (Sigma, Berlin).

Rosen, R. (1991). *Life itself* (Columbia University Press, New York).

Rosen, R. (2012). *Anticipatory systems*, 2nd Ed. (Springer, New York).

Roßnagel, A., Wedde, P., Hammer, V., Pordesch, U. (1989). *Die Verletzlichkeit der Informationsgesellschaft* (Westdeutscher Verlag, Opladen).

Russell, P. (1983). *The Global Brain. Speculations on the evolutionary leap to planetary consciousness* (Turcher, Los Angeles).

Sandkühler, H.J. (1990). *Europäische Enzyklopädie zu Philosophie und Wissenschaften* (Felix Meiner, Hamburg).

Sandkühler, H.J. (1999). *Enzyklopädie Philosophie* (Felix Meiner, Hamburg).

Savranskaya, S., Blanton, T. (2017). NATO expansion: what Gorbachev heard, *National Security Archive*, https://nsarchive.gwu.edu/briefing-book/russia-programs/2017-

12-12/nato-expansion-what-gorbachev-heard-western-leaders-early (retrieved 31.01.2022).

Schellnhuber, H.J., et al. (2016). The challenge of a 4°C world by 2100, eds. Brauch, H., et al., *Handbook on sustainability transition and sustainable peace* (Springer, Cham), https://doi.org/10.1007/978-3-319-43884-9_11 (retrieved 31.01.2022).

Schellnhuber, H.J., et al. (2019). Climate tipping points – too risky to bet against, *nature*, 575, pp. 592-595.

Schnädelbach, H. (1992). *Zur Rehabilitierung des animal rationale* (Suhrkamp, Frankfurt am Main).

Senge, P.M. (1990). *The fifth discipline: the art and practice of the learning organization* (Doubleday, New York).

Senghaas, D. (2007). Abschreckung nach der Abschreckung, *Blätter*, 7, pp. 825-835.

Sharp, G. (2010). *From dictatorship to democracy. A conceptual framework for liberation* (The Albert Einstein Institution, Boston).

Simanowski, R. (2020). *Todesalgorithmus. Das Dilemma der künstlichen Intelligenz* (Passagen Verlag, Wien).

Spiekermann, S. (2021). The value-based engineering handbook, chapter 2, preprint.

Stanley, A. (1999). *Visioneering: God's blueprint for developing and maintaining vision* (Multnomah, Sisters, Oregon).

Stanzl, E. (2017). Maschinen im moralischen Dilemma, *Wiener Zeitung*, July 6, 2017, https://www.wienerzeitung.at/nachrichten/wissen/mensch/903014_Maschinen-im-moralischen-Dilemma.html (retrieved 31.01.2022).

Stauffacher, D., Drake, W.J., Currion, P., Steinberger, J. (2005). *Information and communication technology for peace: the role of ICT in preventing, responding to and recovering from conflict* (Depository Libraries Pilot Project, UN).

Stegemann, B. (2021). *Die Öffentlichkeit und ihre Feinde* (Klett-Cotta,Stuttgart).

Stewart, J.E. (2014). The direction of evolution – the rise of cooperative organization, *Biosystems*, 123, pp. 27-36.

Stonier, T. (1992). *Beyond information: the natural history of intelligence* (Springer, Berlin).

Suchman, L. (2007). *Human-machine reconfigurations – plans and situated actions* (Cambridge University Press, Cambridge).

Teilhard de Chardin, P. (1961). *Die Entstehung des Menschen (Le groupe zoologique humain)* (Beck, München).

Teilhard de Chardin, P. (1964). *Auswahl aus dem Werk (La vision du passé)* (Walter, Olten and Freiburg).

Thatcher, M. (1987). Interview, *Women's Own magazine*, October 31. Reprint, *The Sunday Times*, https://briandeer.com/social/thatcher-society.htm (retrieved 31.01.2022).

The IEEE Global Initiative on Ethics of Autonomous and Intelligent Systems (2019). *Ethically aligned design: A vision for prioritizing human well-being with autonomous and intelligent systems*, 1st ed. (IEEE, Piscataway), https://standards.ieee.org/industry-connections/ec/ead1e-infographic.html (retrieved 31.01.2022).

The Metasystem Transition, http://pespmc1.vub.ac.be/MST.html (retrieved 24.02.2020).

Thurner, S. (2020). *Die Zerbrechlichkeit der Welt* (edition a, Wien).

Tomasello, M. (1999). *The cultural origins of human cognition* (Harvard University Press, Cambridge).

Tomasello, M. (2008). *Origins of human communication* (MIT Press, Cambridge).

Tomasello, M. (2009). *Why we cooperate* (MIT Press, Cambridge).

Tomasello, M. (2014). *A natural history of human thinking* (Harvard University Press, Cambridge).

Tomasello, M. (2016). *A natural history of human morality* (Harvard University Press, Cambridge).

Tomasello, M. (2019). *Becoming human: a theory of ontogeny* (The Belknap Press of Harvard University Press, Cambridge).

Trappl, R. (ed.) (2015): *A construction manual for robots' ethical systems* (Springer, Cham).

Tudyka, K.P. (1973). *Kritische Politikwissenschaft* (Kohlhammer, Stuttgart).

UNCTAD (n.a.). Digitalization offers great potential for development, but also risks, https://sdgpulse.unctad.org/ict-development/ (retrieved 31.01.2022).

Vandana Shiva (2019). *Oneness vs the 1% – shattering illusions, seeding freedom* (New Internationalist Publications, Oxford).

Vernadskij, V.I. (1997). *Der Mensch in der Biosphäre. Zur Naturgeschichte der Vernunft* (Peter Lang, Wien).

Vienna Manifesto on Digital Humanism. Available online: https://www.informatik.tuwien.ac.at/dighum/index.php (retrieved 25 August 2021).

Vine, D. (2015). Garrisoning the globe: how U.S, military bases abroad undermine national security and ham us all, *TomDispatch*, https://tomdispatch.com/david-vine-our-base-nation/ (retrieved 31.01.2022).

Virilio, P. (2000). *The information bomb* (Verso, London).

Wade, N. (2021). The origin of COVID: did people or nature open Pandora's box at Wuhan? *Bulletin of the Atomic Scientists*, https://thebulletin.org/2021/05/the-origin-of-covid-did-people-or-nature-open-pandoras-box-at-wuhan/ (retrieved 31.01.2022).

Wagner, P. (2016). Techno Sapiens: Die Zukunft der Spezies Mensch, documentary, broadcast on 16 November 2016, 3sat, https://www.3sat.de/wissen/wissenschaftsdoku/techno-sapiens-die-zukunft-der-spezies-mensch-100.html (retrieved 31.01.2022).

Wallerstein, I. (1988). *One World, Many Worlds* (Rienner, New York).

Walloth, C. (2016). *Emergent nested systems. A theory of understanding and influencing complex systems as well as case studies in urban systems* (Springer, Switzerland).

Wan, P.Y.-Z. (2011). *Reframing the social. Emergentist systemism and social theory* (Ashgate, Farnham, UK).

Ward, C., Voas, D. (2011). The emergence of conspirituality, *Journal of Contemporary Religion*, 26, 1, pp. 103-121.

Weingartner, P. (1971). *Wissenschaftstheorie I* (Frommann Holz, Stuttgart).

Welsch, W. (1999). Transculturality – the puzzling form of cultures today, eds. Featherstone, M., Lash, S., *Spaces of culture: city, nation, world* (Sage, London).

Wilson, E.O. (1998). *Consilience: the unity of knowledge* (Knopf, New York).

Wimmer, W. (2022). Das gemeinsame Haus Europas sollte das Ziel sein, *Nachdenkseiten*, https://www.nachdenkseiten.de/?p=79607 (retrieved 31.01.2022).

Winiwarter, V. (2021). “Wir müssen anders können”, *Falter*, 51, pp. 65-68.

Wright, E.O. (2010). *Envisioning real utopias* (Verso, London).

Zhao, T. (2019). *Redefining a philosophy for world governance* (Springer Nature, Singapore).

Zhao, T. (2020). *Alles unter dem Himmel. Vergangenjeit und Zukunft der Weltordnung* (Suhrkamp, Berlin).

Zimmermann, R.E. (2012). Naturallianz, Allianztechnkik, eds., Dietschy, B., Zeilinger, D., Zimmermnn, R.E., *Blochwörterbuch* (De Gruyter, Berlin), pp. 349-360.

Zimmermann, R. E. (2014). *H NEA ΠΟΛΥ: Neue Stadtbegriffe auf dem Weg in die Heimat* (LIT Verlag, Berlin).

Zimmermann, R. E. (2015). Mesógios - Zur Struktur der Polis-Netzwerke, eds. Faber, R., Lichtenberger, A., *Ein pluriverses Universum: Zivilisationen und Religionen im antiken Mittelmeerraum* (Fink, Paderborn), pp. 113-130.

Zuboff, S. (2019). *The age of surveillance capitalism. The fight for a human future at the new frontier of power* (Public Affairs, New York).

Index

Printed in the United States
by Baker & Taylor Publisher Services